Advances in Food Protection

NATO Science for Peace and Security Series

This Series presents the results of scientific meetings supported under the NATO Programme: Science for Peace and Security (SPS).

The NATO SPS Programme supports meetings in the following Key Priority areas: (1) Defence Against Terrorism; (2) Countering other Threats to Security and (3) NATO, Partner and Mediterranean Dialogue Country Priorities. The types of meeting supported are generally "Advanced Study Institutes" and "Advanced Research Workshops". The NATO SPS Series collects together the results of these meetings. The meetings are co-organized by scientists from NATO countries and scientists from NATO's "Partner" or "Mediterranean Dialogue" countries. The observations and recommendations made at the meetings, as well as the contents of the volumes in the Series, reflect those of participants and contributors only; they should not necessarily be regarded as reflecting NATO views or policy.

Advanced Study Institutes (ASI) are high-level tutorial courses to convey the latest developments in a subject to an advanced-level audience

Advanced Research Workshops (ARW) are expert meetings where an intense but informal exchange of views at the frontiers of a subject aims at identifying directions for future action

Following a transformation of the programme in 2006 the Series has been re-named and re-organised. Recent volumes on topics not related to security, which result from meetings supported under the programme earlier, may be found in the NATO Science Series.

The Series is published by IOS Press, Amsterdam, and Springer, Dordrecht, in conjunction with the NATO Emerging Security Challenges Division.

Sub-Series

A.	Chemistry and Biology	Springer
B.	Physics and Biophysics	Springer
C.	Environmental Security	Springer
D.	Information and Communication Security	IOS Press
E.	Human and Societal Dynamics	IOS Press

http://www.nato.int/science
http://www.springer.com
http://www.iospress.nl

Series A: Chemistry and Biology

Advances in Food Protection

Focus on Food Safety and Defense

edited by

Magdy Hefnawy
Ag-Tech International Inc.
Greeneville, Tennessee, USA

Springer

Published in Cooperation with NATO Emerging Security Challenges Division

Proceedings of the NATO Advanced Research Workshop on
Advances in Food Security and Safety against Terrorist Threats
and Natural Disasters
Cairo, Egypt
13–15 April 2010

Library of Congress Control Number: 2011924682

ISBN 978-94-007-1102-0 (PB)
ISBN 978-94-007-1099-3 (HB)
ISBN 978-94-007-1100-6 (e-book)
DOI 10.1007/978-94-007-1100-6

Published by Springer,
P.O. Box 17, 3300 AA Dordrecht, The Netherlands.

www.springer.com

Printed on acid-free paper

Preface

Countries and their relevant government agencies, industries, and defense organizations have now recognized that the protection from deliberate contamination of the entire food supply chain, by terrorists or criminals, is absolutely essential for the safe delivery of food and ultimate food security. The NATO Public Diplomacy Division Science for Peace and Security Section (SPS) sponsored an Advanced Research Workshop on food safety and security against terrorist act or natural disaster. The workshop served its purpose for scientists from a wide span of countries to set a foundation for dealing with the global issue of advancing food security and safety. The three main objectives of the workshop held April 13–15, 2010, in Cairo, Egypt, were the following:

1. Allow scientists and industry experts from various countries to make a critical assessment of existing knowledge on food security and safety as defense against terrorist attacks.
2. Develop a strategy to counter the risks against food supply from terrorist acts or natural disasters.
3. Identify directions for future research and approaches to strengthen prevention and responses to food terrorism.

Food Security, which can be defined as having access to sufficient and safe food supplies, can be threatened by contamination at every level of the supply chain from initial food production and distribution through to the consumption of the food. In the initial section of the workshop, challenges to food safety and security were addressed to define the scope of the problems all nations face and to set the stage for dealing with contamination of the food supply by either environmental causes or deliberate interventions aimed at impacting both the physical health of a population and the economic health of a country. Leading experts defined threats from microorganisms as well as chemical and radiological toxic contaminants. Therefore, the participants then presented a series of papers to discuss and define preventative measures and strategies to help participants and readers of this book build a framework to help nations cope with diverse challenges to the complex global food system. Measurement of impact of intentional contamination can be made with risk or threat modeling programs, and these models can also be used to test the effectiveness of various proposed interventions. Preparation for dealing with a crisis in the food supply from both governmental and food industry perspectives was a major theme in several of the papers presented. Prevention of contamination is certainly a major goal of any food safety or food defense program, and common methodologies may be employed. This can occur once both industry and supporting governmental agriculture and regulatory agencies come to recognize both the scope of the problems and the emerging set of tools from advanced technologies available to detect problems at early stages and then to employ effective means to eliminate potential problems. The lessons learned from the deliberate contamination of food protein ingredients with melamine point out the need for all countries to develop the ability to have data bases for food ingredients and to have food producers and shippers develop systems to trace the path of food ingredients in the logistics chain. Participants in the workshop felt that investments do need to be made to strengthen the food/agriculture infrastructure by implementing food security plans that would use principles set up by hazard/risk analyses with critical control points. The reduction of waste and spoilage in the food distribution chain could largely offset implementation costs.

Several presentations in the final session of the workshop dealt with emerging technologies to improve food safety and security. Improved methodologies for both detection of contaminants and control or inactivation of contaminants may help improve both food safety and security, but implementation on a systematic basis will require concerted efforts on the part of both industry and governments with the assistance from international agencies to communicate pathways for international collaborations. This is required to facilitate the most appropriate actions that both developed and developing nations should take to move into a mode of crisis prevention as well as preparation for better responses to crises that will still emerge. There were several specific recommendations that emerged from the workshop:

1. To establish a "Center of Excellence for Food Safety" affiliated with a local university in the Mediterranean area with links to similar centers in the United States or NATO partner countries.
2. To create an active international networking body from academia, the food industry, and government agencies to support national food safety and defense planning and research. Some existing international professional societies working in conjunction with the World Health Organization and the United Nations Food and Agriculture Organization could facilitate such an undertaking by linking information technology resources.
3. To develop a strategy to increase the awareness of food protection and defense for the general populace as well as those in agriculture and the food industry with the help of those international agencies. It can be stated that some rather simple changes instituted at the local producer level can mitigate the need for more extensive changes higher up the food chain.

This workshop did set in place some means to consider for improving lines of communication among different countries by helping identify a number of different resources. The published proceedings will serve as one such resource, but the human resources brought together in this workshop also need to continue to reach out to others to grow both the awareness of the need to protect the food production enterprise and also to communicate some of the tools useful to improve the delivery of safe food products around the world.

The organization committee of the workshop would like to thank NATO Public Diplomacy Division, Science for Peace and Security Section (SPS) for the support and encouragement, especially to Dr. Deniz Beten, Dr. Susanne Michaelis, and Ms. Alison Trapp.

C. Patrick Dunne
Magdy Hefnawy

Contents

Contributors

Ingrid Aguiló-Aguayo
Department of Food Technology, University of Lleida, Avda. Alcalde Rovira Roure 191, Lleida 25198, Spain

Hami Alpas
Department of Food Engineering, Middle East Technical University, Ankara 06531, Turkey
imah@metu.edu.tr

Slavko Bogdanovic
The Business Academy, Independent and Non-State University of Novi Sad, Novi Sad, Serbia
slavkob@nadlanu.com

Faruk Bozoğlu
Department of Food Engineering, Middle East Technical University, Ankara 06531, Turkey
bozoglu@metu.edu.tr

Francis (Frank) F. Busta
National Center for Food Protection and Defense, University of Minnesota, St. Paul, MN 55108, USA
fbusta@umn.edu

Thomas A. Butterworth
Austin Food Tech Inc., 9 Yemen Street #30, Mohandessin, Giza 12411, Egypt
foodtech@alum.mit.edu

J. Claude Cheftel
Université des Sciences & Techniques, F-34000 Montpellier, France
claude.j.cheftel@orange.fr

C. Patrick Dunne
Combat Feeding Directorate, U.S. Army Natick Soldier Research Development and Engineering Center, Natick MA 01760, USA
c.patrick.dunne@us.army.mil

Pedro Elez-Martínez
Department of Food Technology, University of Lleida, Avda. Alcalde Rovira Roure 191, Lleida 25198, Spain

Khalil I. Ereifej
Department of Nutrition and Food Technology, Faculty of Agriculture, University of Science and Technology, 3030, Irbid 22110, Jordan
ereifej@just.edu.jo

Magdy Hefnawy
Ag-Tech International, Inc., 237 Cal Dobson Trail, Greeneville, TN 37743, USA
magdyh@centurylink.net

Shaun P. Kennedy
National Center for Food Protection and Defense, University of Minnesota, St. Paul, MN 55108, USA
kenne108@umn.edu

Sameeh A. Mansour
Environmental Toxicology Research Unit (ETRU), Department of Pesticide Chemistry, National Research Centre, Tahrir Street, Dokki, Cairo, Egypt
samansour@hotmail.com

Olga Martín-Belloso
Department of Food Technology, University of Lleida, Avda. Alcalde Rovira Roure 191, Lleida 25198, Spain
omartin@tecal.udl.cat

Radomir Radovanovic
Department of Food Safety and Quality Management, Faculty of Agriculture, University of Belgrade, 11080 Belgrade – Zemun, Nemanjina 6, Belgrade 11080, Serbia
laler@agrif.bg.ac.rs

Hamed M. Roushdy
Department of Radiation Biology, National Center for Radiation Research and Technology, Atomic Energy Authority, Nasr City, Cairo, Egypt
roushdy@excite.com

Madeleine Smith
Chemical Engineering, University of Birmingham, Edgbaston, B15 2TT Birmingham, UK
m.smith.2@bham.ac.uk

Robert Soliva-Fortuny
Department of Food Technology, University of Lleida, Avda. Alcalde Rovira Roure 191, Lleida 25198, Spain
rsoliva@tecal.udl.cat

Dina Solodoukhina
Department of Public Health, Kursk State Medical University, Karl Marx str. 3, Kursk, Russia
solodin_kursk@mail.ru

Chapter 1
NATO-SPS Pilot Study on Food Chain Security: Findings and Recommendations

Hami Alpas and Madeleine Smith

Abstract The chapter reports the outcomes of an ongoing NATO-SPS Pilot Study on "Food Chain Security." These include (1) Overview of the food system in participating countries; (2) Prevention, surveillance and detection systems and (3) Response systems. A survey was carried out in order to identify the concerns of the food industries with regard to contamination of the food chain (i.e. what did the industry perceive as the main hazards associated with the food chain?); to evaluate the extent to which food safety management systems had been implemented as a safeguard against contamination incidents and to determine the feasibility and limitations of conducting such a survey within participating countries. A survey tool was prepared and sent to public and private food sector(s) by each group member. A total of 18 different countries participated in the pilot study and nine countries completed the survey. Participants' views were gathered, and the outcomes are discussed in terms of reducing possible risks and threats to the food system. The detailed analysis of the outcomes will be a key to address to the countermeasures that are necessary to be taken to minimize the adverse effect(s) on human health, society, and environment at large. The possible scenarios developed by the experts in the project duration are shared. These are believed to be useful in developing a prototype system and for sharing information related to acts of terrorism to the environment, agriculture, and water systems. The final outputs given here allow some comparison between country partners and identify common weaknesses of the food systems in general. The importance of continuing exchange of experience between participating countries in the area of legal regulations, logistical and institutional aspects of food safety is highly encouraged.

Keywords Food chain security • NATO pilot study findings

H. Alpas (✉)
Department of Food Engineering, Middle East Technical University, Ankara 06531, Turkey
e-mail: imah@metu.edu.tr

M. Smith
Chemical Engineering, University of Birmingham, Edgbaston, B15 2TT Birmingham, UK
e-mail: m.smith.2@bham.ac.uk

M. Hefnawy (ed.), *Advances in Food Protection*, NATO Science for Peace and Security Series A: Chemistry and Biology 1, DOI 10.1007/978-94-007-1100-6_1,

1.1 Introduction

Food chain security can be breached by contamination at any point in the food chain. Using the FAO definition of food chain security:

> Food security exists when all people at all times have physical and economic access to sufficient, safe, and nutritious food to meet their dietary needs and food preferences for an active and healthy life style [1].

A contamination incident that makes the food unsafe will breach the security of the food chain. Such contamination incidents have been identified and recorded and can be accidental or deliberate. The vast majority is accidental and include microbial, physical, and chemical contamination. Contamination incidents occur accidentally during food production and include physical contaminants such as glass, metal, or plastic and microbial contamination by pathogens [2]. Chemical contamination may also occur but appears to be less common [3]. Microbial contamination is well documented and can result in major outbreaks e.g. [4]. Occasionally a deliberate contamination incident occurs. These are typically carried out for commercial gain, for example the adulteration of cooking (rapeseed) oil with aniline in Spain in 1981 [5, 6]. Also very occasionally a deliberate contamination incident can be attributed to political or social concerns [7].

In many countries the food industry implements food safety management systems to minimize the number of accidental contamination incidents. These safety systems may be proactive (e.g. HACCP based systems) and reactive (e.g. Recall systems). Although designed to address the issues of accidental contamination, these systems tend to also offer some protection against deliberate contamination incidents.

The interrelationship between food safety and security measures in protecting the food supply has been recognized by both the World Health Organization (WHO) and United States Food and Drug Administration (US FDA).

The WHO in its report on "Terrorist Threats to Food" [8] noted that "outbreaks of both unintentional and deliberate food borne disease can be managed by the same mechanism. The key to preventing food terrorism is establishment and enhancement of existing food safety management programmes and implementation of reasonable security measures. Prevention is best achieved through a cooperative effort between government and industry, given that the primary means for minimizing food risks lie with the food industry."

During the pilot study on food chain security, a survey was carried out by committee participants to examine the matter of food contamination and the associated control mechanisms in their own countries. This chapter summarizes and discusses the results of this survey.

1.2 Survey

The contamination and disruption of components of the food supply system may have adverse effects on:

- public health through injury or death;
- the viability of the food supply;
- public confidence in the safety of the food we eat; and
- social and political instability.

Therefore, the safety and security of the food supply is a priority in any comprehensive counter terrorism strategy, and threats to the safety of the food supply are already a reality. Remembering that in the modern food supply system:

- many of the agricultural materials used by processors are seasonal and are held either in bulk storage or as processed product;
- the processing steps generally, but not always, control or remove dangerous organisms and contaminants;
- food is generally processed as branded products in large scale plants;
- food is moved through the supply chain as quickly as possible with minimal inventory, particularly of short-shelf life products, to ensure low, internationally competitive prices to consumers;
- an increasing proportion of fresh food produced in large scale plants and is sold through major retail chains;
- a significant proportion of food is sold directly to the consumer through restaurants and other food service outlets; and
- water is a significant ingredient throughout the food chain – in irrigation, processing, and as an ingredient.

It would not be wrong to claim that "food is both a potential victim of direct contamination and a likely casualty of disruption to basic infrastructure services" [9].

The main objectives of the survey were:

- To identify the concerns of the food industries with regard to contamination of the food chain (i.e. what did the industry perceive as the main hazards associated with the food chain?)
- To evaluate the extent to which food safety management systems had been implemented as a safeguard against contamination incidents
- To determine the feasibility and limitations of conducting such a survey within participating countries.

A total of 15 questions were prepared (Table 1.1).

1.3 Methodology

The survey questions had categorical answers providing nominal data. The survey was to be carried out in several countries and in a variety of languages. Each country conducted the survey at a different time. Belgium and Finland carried out their surveys in 2004, other countries later and the last were Portugal and Russia in 2008.

Table 1.1 Survey questions

Box. 1

Participants information

Identification questions:

1) Please identify which sector of food chain you work in.

Agriculture
Transport
Production
Depositing
Distribution

Definition:

- Agriculture: raw material conceming farms, feed industry, farming.
- Transportation 1: from feed producer to farm
- Transportation 2: from farm to the processing
- Transportation 3: from processing to distributor
- Processing: all steps of transformation of raw material into consumable products
- Storage: silos, cool-houses.
- Distribution: from distributor to retail shop

2) What kind of products do you produce?
   1. Milky
   2. Meat
   3. Cereals
   4. Vegetables
   5. Fruits
   6. Drinks
   7. Others (please specify) ....

Questionnaire

1) Which of the followings would be the most important risks for your product:
   a) chemicals,
   b) bacteria/virus
   c) radiation
   d) energy crisis
   e) water crisis
   f) raw material
   g) GMO
   h) Others…(please, specify)
2) Do you have internal discussion in the company about terrorist attacks?
3) Do you regularly check your employee records?
4) Do you have a food safety monitoring program?
5) Do you carry out any one of this inspection before you accept the raw materials:
   a) Chemicals
   b) Microbiological
   c) Radioactivity
6) About your suppliers:
   a) Domestic
   b) International

(continued)

Table 1.1 (continued)

7) About your company's holders.
   a) Multinational
   b) National
8) Who is in charge of the food distribution?
   a) Yourselves
   b) Your contractor
9) Who is in charge of food storage?
   a) Yourselves
   b) Your contractor
10) How quick do you think that you can respond to a food safety problem?
   a) immediately
   b) within a few hours
   c) a day
   d) can not tell
11) Do you have crisis management system?
   a) yes
   b) no
12) Which of the followings is more vulnerable from the point of food safety?
   a) Raw material
   b) Processing systems
   c) Storage systems
   d) Distribution systems
   e) Retail shops
13) Do you think that GMO is a safety concern in your production system?
   a) Yes
   b) No
   c) No idea
14) Do you have a recall system for the products?
   a) Yes
   b) No
15) Do you have tested your recall system?
   a) Yes
   b) No

Thank you for being involved!

Participating countries differ in the structure of their food industries and control mechanisms. As a consequence the sampling frame and data collection methods also varied. This combined with the data type and timeframe, made it impossible to standardize the data collection in such a way that any quantitative statistical analysis could be applied. The data has, therefore, been analyzed descriptively and is presented as percent responses by country.

1.4 Results

1.4.1 Profile of the Respondents

Nine countries carried out the survey. The survey asked respondents to indicate which sector of the food chain they occupied and were given the options of Agriculture, Transportation, Processing, Storage and Distribution. Some countries were able to include all sectors in the survey; some only covered a subgroup. Some recipients fitted into more than one category. It should be noted that the respondents were all producers/manufacturers/distributers. None of the samples included food service or catering establishments. Respondents were also asked to identify the food groups they handled and were given the options of Dairy, Meat, Cereals, Vegetables, Fruits, Beverages, and Other (see Table 1.2). The categories covered depended on the products popular in the participating countries and the sample frame, but at least one of the high-risk dairy and/or meat categories were covered by all participants who answered this section. Russia did not specify the categories covered. Other respondents included bakery goods, processing aids, snack foods, nuts, pastry, sauces, vegetarian ready meals, oil, confectionary, convenience foods, fish and fishery products, and vinegar.

1.4.2 Food Chain

Respondents were asked about their suppliers. Respondents had both domestic and international suppliers. All respondents used some international suppliers, and the percentage ranged from 71% (Turkey) to 4% (UK). In the UK, 68% answered that both types of supplier were used.

Respondents were asked whether they kept control of and responsibility for distribution and storage of product. Between 72% (UK) and 82% (Czech R) were responsible for food storage themselves while between 18% (Czech R) and 47% (T, UK, Bu) distributed the food themselves. Other respondents contracted out or shared responsibility for storage and distribution with other companies. Between 30% (UK) and 82% (T) contracted out and between 18% (Czech R) and 23% (UK) shared responsibility. Results add up to more than 100% as some respondents used more than one option.

Overall the responses indicate a complex food chain in all participating countries, involving both domestic and international suppliers and using multiple companies to produce, store, and distribute the food.

Table 1.2 Profile of respondents

Country	No. premises surveyed	No responses (n)	% response	Sample method	Language	Distribution of respondents
Turkey	25	17	68%	Post, fax and Email	English	15 Agriculture, 2 Distribution
Bulgaria	Not stated	15	Not stated	Not stated	Not stated	All categories, 80% processors
UK	250	47	19%	(internet) Postal Respondent completed	English	All categories except agriculture. 100% processors
Czech rep.	Not stated	30	Not stated	Not stated	Not stated	All categories 30% processors
Finland	Not stated	17	Not stated	Not stated	Not stated	All categories 100% processors
Belgium	Not stated	132	Not stated	Respondent completed via email and conventional mail	Dutch and French	All categories 63% processors
Romania	30	27	90%	Opportunistic sample personal delivery	Romanian	All except agriculture
Portugal	14	14	100%	Opportunistic Telephone and email	Portuguese	All categories 57% processors
Russia	270	65	24%	Email and personal delivery	Russian	Not stated

1.4.3 *Hazards and Risks*

Respondents were asked to identify which hazards they considered constituted the most significant food safety risk to their process. They were given the options of chemical contamination, bacteria/viral contamination, radiation, energy crisis, water crisis, raw material, GMO and other. Chemical and microbiological contamination and raw materials were the most frequently identified (Fig. 1.1)

- Chemical 12% (Czech R) – 80% (B & UK)
- Microbiological 6% (T) – 76% (UK)
- Raw materials 13% (Czech R & Bu) – 72% (UK)

Of least concern were radiation, energy crisis, water crisis, and GMO.

- Radiation was identified by four countries as a consideration – 12% (Turkey); 18% (Russia); 20% (Bulgaria); 29% (Finland) and all others 0%.
- An energy crisis was identified in all countries as a concern but by limited numbers of respondents 7%(Portugal) – 53% (Finland)
- A water crisis was identified in all countries as a concern but by limited numbers of respondents 13%(Bulgaria & Czech R) – 59% (Romania)

Five countries contained respondents who were concerned about GMO's constituting a food safety risk. These were UK (48%), Russia (35%), Belgium (22%), Turkey (18%) and Portugal (14%). For all other countries the response rate was 0% (Fig. 1.2).

Respondents identified the contamination of raw materials as a significant concern. When asked to identify which point in the food chain they felt was most vulnerable, they again identified raw materials as being of high concern. The options given were Raw material supply, Processing systems, Storage systems, Distribution systems, Retail shops. Raw materials were identified as the most vulnerable point by all countries except Bulgaria where it was equal with processing (47%) (Fig. 1.3).

In summary, the main concerns expressed by the respondents are the traditional hazards of chemical and bacterial contamination. Contamination by radiation or

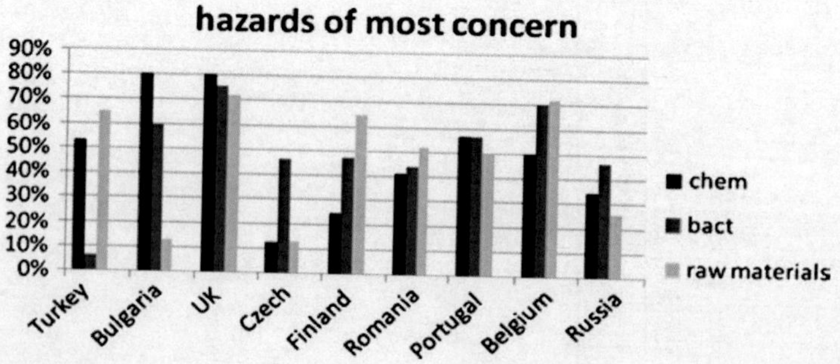

Fig. 1.1 Hazards of most concern

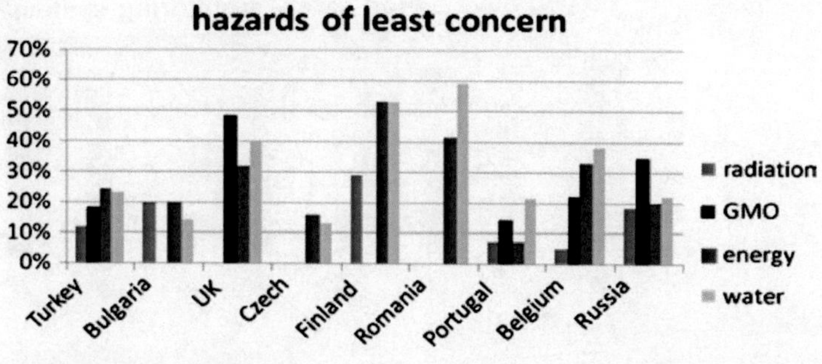

Fig. 1.2 Hazards of least concern

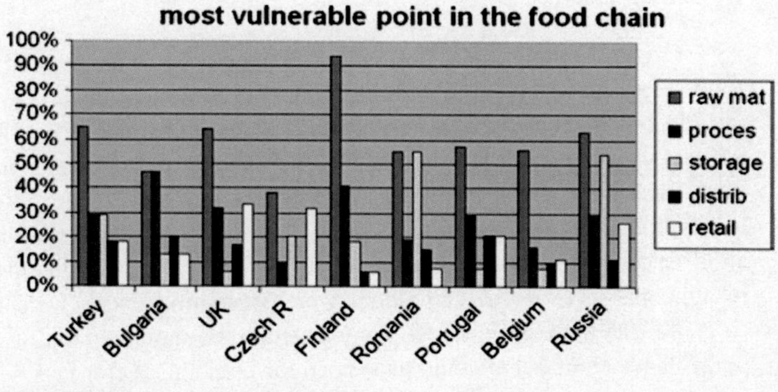

Fig. 1.3 Most vulnerable point in the food chain

by GMO's was not identified as common concerns. Some concern was expressed in all countries about the interruption of energy or water supply, but only in Romania and Finland did this reach more than 50%. In all countries the respondents identified the acquisition of raw materials as being the most vulnerable point in the food chain.

1.4.4 Management of Food Chain Security

The respondents were asked questions about their management of food chain security. The questions covered issues such as food safety monitoring, recall, and crises management as well as whether they had discussions about terrorist attacks or carried out security checks.

Respondents were asked if they had a food safety monitoring system. In some countries, if they answered 'yes' they were also asked what type (e.g. HACCP, GMP). The majority of respondents answered yes, that they had a food safety monitoring system. In some countries 100% had such a system (Fig. 1.4). Where respondents were asked, the most common system used was HACCP (100% UK,

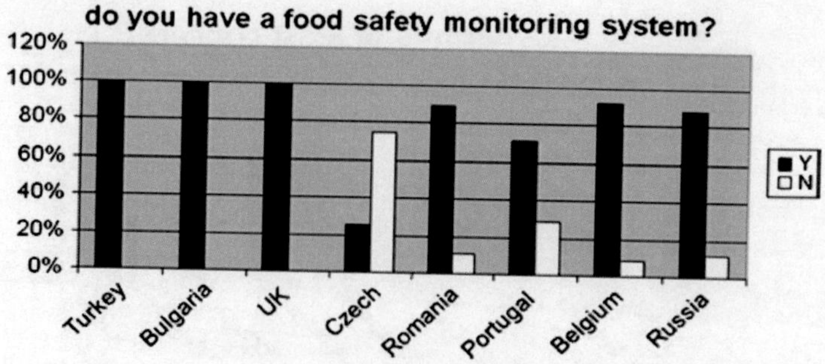

Fig. 1.4 Do you have a food safety monitoring system?

25% Czech Republic, 48% Romania). It should be remembered, however, that a food safety management system has been a requirement under EC law since 1995 (article 2 of Directive 43/93, subsequently replaced by article 5 of Regulation (EC) 852/2004) so it would be expected that the majority of companies in member states would have one. Food businesses outside the EU must meet these legal requirements if wishing to trade with member states, so the EC legislation may have had an effect on non member states as well. It should also be remembered that all the respondents were in food production or distribution. None were food service or caterers. Had these groups been represented in the sample group, it is likely that the percentages complying would have been lower as this sector lags behind the manufacturing sector in its uptake of HACCP and other safety management systems [10]. The widespread implementation of food safety management systems is significant as they assist in the production of safe food, helping to prevent food chain contamination and preserve food security.

Food safety management systems are a proactive way of managing food contamination. Reactive systems such as product recall and crisis management will also be required to ensure food chain security. Respondents were also asked about these mechanisms.

In all countries some respondents had a recall system in place. This ranged from 100% of respondents in Finland to 50% in the Czech Republic (Fig. 1.5). Respondents were also asked if they had used or tested these systems. The response varied according to country. In Finland and the UK, 94% of the respondents who had a recall system had used or tested it. In Turkey it was 86%, in Portugal 80%, Romania 74%, Russia 55%, Czech Republic 50%, and Bulgaria 13%.

In all countries some respondents had a crisis management system in place. This ranged from 57% of respondents in Portugal to 100% in Belgium. In other countries between 60% and 91% of the respondents had such systems.

Respondents were also asked about the speed with which they could respond to a food safety problem. Many respondents in countries with extensive management and recall systems were confident about being able to respond quickly (either immediately or within a few hours) (see Table 1.3).

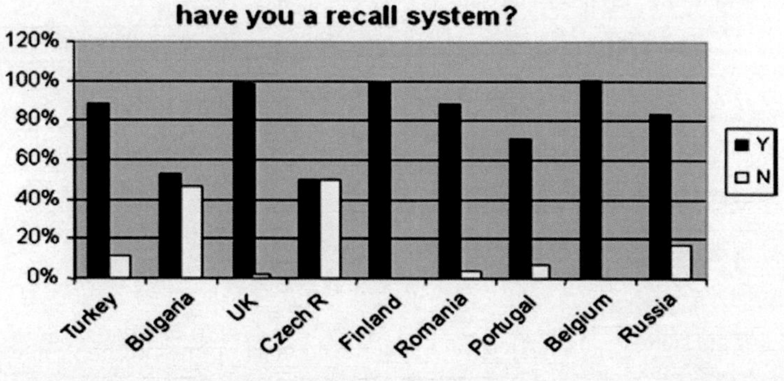

Fig. 1.5 Have you a recall system?

Table 1.3 Expected response times for a food safety problem

Q10	Turkey	Bulgaria	UK	Czech rep.	Russia	Romania	Portugal	Belgium
Immediately	53%	13%	61%	13%	51%	70%	7%	20%
Few hours	29%	7%	36%	40%	26%	30%	50%	46%
A day	12%	47%	4%	13%	9%	3%	14%	19%
Don't know	6%	33%	0%	34%	14%	0%	7%	12%

Table 1.4 Respondents having formal discussions or meetings about terrorist activity

Q2	Turkey	Bulgaria	UK	Czech rep.	Russia	Romania	Portugal	Belgium
Y	53%	27%	8%	15%	60%	0%	36%	28%
N	47%	73%	92%	85%	40%	100%	36%	72%

The management systems referred to above is mainly in place to prevent accidental contamination of the food chain. Respondents were also asked if they carried out specific actions related to deliberate or terrorist activity. When asked if they ever had meetings or formal discussions in the company about terrorist attacks, very few respondents answered 'yes'. Only in Turkey and in Russia did over half the respondents carry out such meetings and in most other countries this was reduced to a very small number (see Table 1.4).

Respondents were also asked about checks carried out on staff. A high proportion of the respondents claimed to do this, but unfortunately the wording of the question was somewhat misleading on some questionnaires and is likely to be measuring very different aspects in different countries (Table 1.5).

In some situations this question appears to have been misinterpreted to ask whether competency references are checked, rather than security references. However, it can be identified that in all countries except the UK (Finland missing) the majority of respondents check employee records, although for what is not clear.

Table 1.5 Respondents claiming to check Employee records

Q3	Turkey	Bulgaria	UK	Czech rep.	Russia	Romania	Portugal	Belgium
Y	94%	100%	36%	90%	89%	100%	71%	31%
N	6%	0%	64%	10%	11%	0%	28%	69%

1.5 Discussion

This survey was carried out using different sampling protocols and methods in each country. Most of the sampling protocols were opportunistic, and the respondents were all volunteers. Some sectors of the food chain are over represented (manufacturers), and others are underrepresented or absent (catering and food service). The results cannot therefore be interpreted as being wholly representative of the food industry in any country, and conclusions must be considered in the context of these confounders. Should any further action be taken as a result of this survey, it is recommended that the outcomes be re-evaluated in the relevant individual country using more reliable sampling protocols.

Overall, the responses indicate a complex food chain involving both domestic and international suppliers and using multiple companies to produce, store, and distribute the food. According to DEFRA [11] the use of international suppliers will act to enhance food security. This is particularly the case for developed countries with good infrastructures and established trading mechanisms. DEFRA consider that the flexibility built into such supply chains provides a certain resilience. Point disruptions of a complex chain may result in temporary and/or local problems of safety or access, but do not usually create a fundamental threat to food security, for example the compromised safety of Irish pork and pork products in 2008 from dioxin contaminated animal feed [12].

The survey shows that many of the respondents are concerned about contamination of raw materials, particularly with chemical and microbial hazards. Many have in place management systems to address these hazards, including HACCP, recall procedures, and crisis management systems. In countries where high proportions of the respondents claimed to have these systems in place, they were also confident about responding quickly to a food problem. It is the policy within EU member states that these management systems, including additional systems such as traceability, should become universal in all food businesses (Regulation (EC) 852/2004; Regulation (EC) 178/2002). It is considered that such systems act as a bulwark against food chain contamination and consequently act to enhance food chain security. In many participant countries they will have been in place for over a decade, and when they are implemented in conjunction with a unified notification, hazard warning, and laboratory system, they can work well to identify and control food safety incidents.

Such management systems are not perfect and can only provide a certain level of protection. It may also be that there are gaps in food chain security not addressed by the management systems even when working well. One very significant gap, the awareness of the impact of an energy crisis, was identified by this survey. Of all the respondents in this survey, only in Finland did more than 50% of the respondents feel energy was a major consideration. In Romania 41% of respondents expressed concern about energy, but in all other countries less than a third felt it to be an issue (see chart 3). Interruption of energy supply may not in itself cause contamination of food, although interruption of the cold chain would significantly increase spoilage and pathogenic growth. If prolonged, however, it would have a major effect on food chain security, interrupting all sectors from production of raw materials through to retail sale, disrupting access as well as production and delivery. This risk seems to be underestimated by all the participants in the survey and may be one area where competent authorities and other agencies would find it appropriate to liaise with industry. Energy supply is a critical issue, and a threat in this area would over ride all the safety measures working on a day to day basis to control contamination incidents. As with traditional contamination, energy interruption could be the result of accidental or deliberate action.

Very few of the respondents had meetings or discussions about terrorist activity (Table 1.4). It may be that the general security arrangements many producers/food business have in place to control more mundane but still illegal activities such as theft and vandalism, are thought to be sufficient to control deliberate attempts at contamination or disruption. Systems that control movement of staff and visitors to ensure control of accidental contamination will also have some effect on controlling attempts at deliberate contamination. WHO recommend that staff are screened, and those with access to critical areas are security checked. The FDA also support staff screening (making the point, however, that it should apply equally to all staff) in their Guidance for Industry [13]. The majority of respondents in the survey claimed to check employee records, although from the wording of the question it is not possible to tell what these checks establish. The area of employee security is a very difficult issue, and the cost (in money and good will) of additional, possibly onerous checks on potential and existing employees would need to be considered against the actual risk within the company. The number of documented, published cases where individual employees have deliberately contaminated food stuffs is very small. Anecdotal cases of such deliberate contamination are more common but are frequently traced back to a disgruntled employee who feels slighted or mistreated by the company [14]. These cases would not be addressed by increased security checking but by better management.

All surveys have inherent confounders related to the mechanism of data collection. Ensuring a representative sample group is a major concern. As mentioned above, the respondents in this survey were all volunteers. The sampling process in many countries was opportunistic. These factors make it impossible to establish how representative the responses are of the country's relevant food sector. Such confounders are common in work of this type but exacerbated by the scale and breadth of this project. The survey also identified further limitations attendant on such an ambitious data collection process. One important limitation is that of language. The original survey was developed in English and was then translated as required. The questions should have been piloted in each country before use to establish that the questions were as consistent as possible between participants. Inconsistency certainly existed as can be identified for example by Q3 which asked about employee checks. It was not clear what checks the respondents actually made or how. Other confounders include differences in sampling method, sample size and type, time of sampling, and method of completion. This variation in methodology between countries severely limited the type of analysis that could be carried out. Given the differences between the countries involved and the non uniformity of the food industry, it is unlikely that these basic parameters could ever be controlled sufficiently to allow more rigorous or comparative analysis on a project of this scale. However, some useful information was gathered regarding the extent to which those volunteering to complete the survey were implementing certain systems and which food safety hazards were considered important. It was also possible to identify areas where further research could be carried out or where competent authorities or governments may wish to consider increased liaison with the food industry.

1.6 Conclusion

The participants in the survey identified the traditional food safety contaminants as their major areas of concern i.e. chemical and microbial contamination of foodstuffs, particularly raw materials. This was found in all countries.

The number of participants reporting the use of food safety management, recall, and crisis management systems varied between countries. Where limited implementation of these systems is reported, this may indicate a lower level of security within the food industry. However, this should be assessed in the context of the country and also considering the limitations of the sampling frame of the survey.

Some areas of importance were identified in the survey that were not being considered by the participants but which would have an impact on food chain security such as the impact of energy supply disruption.

Multinational surveys may provide a useful source of qualitative data on certain subjects, but such data must be considered within the limitations of the survey methodology and in the context of the participating countries. Due to the variation in methodology across participating countries, it is not possible to analyse such data in a statistically rigorous manner or to draw valid comparisons between countries.

Acknowledgements The authors would like to thank to the members of the Pilot Study who have contributed to the survey and collection of data: namely; Momtchil Sidjimov (Bulgaria), Ales Komar (Czech Republic), Kyösti Orre (Finland), Beyazit Çirakoglu and Faruk Bozoglu (Turkey), Alexandria Veiga (Portugal), Liviu Galatchi (Romania) and Irina Ermakova (Russian Fed.). We are grateful for their kindness.

References

1. FAO (1996) Rome declaration on world food security; world food summit 13–17 Nov 1996, Rome. http://www.fao.org/docrep/003/w3613e/w3613e00.HTM
2. FSA (2008a) Food Alerts May – Aug 2008. Hard plastic in apple pies http://www.food.gov.uk/enforcement/alerts/2008/aug/livwellpies; Rubber contamination in crisps. http://www.food.gov.uk/enforcement/alerts/2008/aug/walkers; *Salmonella* in Bovril http://www.food.gov.uk/enforcement/alerts/2008/aug/salmonellainbovril; *Listeria* in cooked turkey. http://www.food.gov.uk/enforcement/alerts/2008/jul/turkey; Plastic contamination in 'Build up' powder; http://www.food.gov.uk/enforcement/alerts/2008/may/buildup; Glass contamination in flapjacks; http://www.food.gov.uk/enforcement/alerts/2008/may/flapjack; Hard plastic in cheese spread. http://www.food.gov.uk/enforcement/alerts/2008/may/primula
3. FSA (2008b) Food alerts taint in Salmon (diesel and white spirit). http://www.food.gov.uk/enforcement/alerts/2008/feb/salmonupdate2; http://www.food.gov.uk/enforcement/alerts/2008/feb/salmon
4. Cowden JM, Ahmed S, Donaghy M, Riley A (2001) Epidemiological investigation of the central Scotland outbreak of *Escherichia coli* O157 infection, November to December 1996. Epidemiol Infect 126(3):335–341
5. Sanchez-Porro Valades P, Posada de la Paz PM, de Andres Copa P, Gimenez Ribota O, Abaitua Borda I (2003) Toxic oil syndrome: survival in the whole cohort between 1981 and 1995. J Clin Epidemiol 56:701–708
6. Tabuenca JM (1981) Toxic allergic syndrome caused by ingestion of rapeseed oil contaminated with aniline. Lancet 2:567–568
7. Török TJ, Tauxe RV, Wise RP, Livengood JR, Sokolow R, Mauvais S, Birkness KA, Skeels MR, Horan JM, Foster LR (1997) A large community outbreak of salmonellosis caused by intentional contamination of restaurant salad bars. JAMA 278(5):389–395
8. WHO (2008) Terrorist threats to food, ISBN 92 4 154584 4. http://www.who.int/foodsafety/publications/general/en/terrorist.pdf. Accessed 30 Sept 2010
9. Wells D (2003) Safeguarding Australia: agri-terrorism and food chain security. http://www.afgc.org.au/cmsDocuments/Food%20Chain%20Security.pdf. Accessed 30 Sept 2010
10. Taylor E, Kane K (2005) Reducing the burden of HACCP on SMEs. Food Control 16:833–839
11. DEFRA (2006) Food security and the UK: an evidence and analysis paper. http://www.defra.gov.uk/science/Project_Data/DocumentLibrary/FT0352/FT0352_4705_FRP.doc
12. FSA (2008c) News release Irish pork update. http://www.food.gov.uk/news/newsarchive/2008/dec/irishpork
13. FDA (2007) Guidance for industry: retail food stores and food service establishments: food security preventive measures guidance. http://www.cfsan.fda.gov/~dms/secgui18.html
14. Anon (2003) Nicotine poisoning after ingestion of contaminated ground beef. MMWR 52(18):413–416. http://www.cdc.gov/mmwr/preview/mmwrhtml/mm5218a3.htm. May 9, 2003

Legislation

Council Directive 94/43/EC Hygiene of Foodstuffs.http://www.okyanusbilgiambari.com/tgm/Eu-lex/FdSafetyDir93_43-HygieneFood.pdf. Repealed 1 Jan 2006

Regulation (EC) no 852/2004. http://www.food.gov.uk/multimedia/pdfs/hiojregulation.pdf

Regulation 178/2002. http://eur-lex.europa.eu/LexUriServ/site/en/oj/2002/l_031/l_03120020201en00010024.pdf

Chapter 2
Food Safety and Bioterrorism from Public Health Perspective

Dina Solodoukhina

Abstract The chapter discusses the issues of food safety and bioterrorism from the position of public health. The author gives the overview of the problem focusing on agricultural sector as a particularly vulnerable element in the farm-to-table food continuum, describes the possible scenarios of bioterroristic attacks and the necessary preventive activities to provide food safety. The article also contains the results of the original study conducted in two Russian regions on the perception of risk related to food safety by consumers.

Keywords Bioterrorism • Food safety • Risk perception • Russia

2.1 Bioterrorism as a Challenge for the Global

In the last few years changes in the political and economic situation in the world make us think about the new threats to the global peace and safety. While all of us consume food products every day that is essential for our life, bioterrorism becomes one of the real possibilities, which must be prevented before we have numerous victims.

Bioterrorism can be defined as terrorism by intentional release or dissemination of biologic agents (bacteria, viruses, or toxins); these may be in a naturally-occurring or in a human-modified form. Still we must mention that Biological and Toxin Weapons Convention in 1972 banned their production and any form of use [1]. Other definitions of bioterrorism include "the unlawful use of viruses, bacteria, fungi, toxins, or other pathogenic material against a government, the civilian population, livestock, crops, or any segment thereof, in furtherance of political,

D. Solodoukhina (✉)
Department of Public Health, Kursk State Medical University, Karl Marx str. 3, Kursk, Russia
e-mail: solodin_kursk@mail.ru

M. Hefnawy (ed.), *Advances in Food Protection*, NATO Science for Peace and Security Series A: Chemistry and Biology 1, DOI 10.1007/978-94-007-1100-6_2,

social and/or economic objectives" [2], or "the use of dangerous biological agents for inflicting damage to the life and health of people in order to reach goals of a political and materialistic nature" [3].

A recent opinion poll on "Attitudes in the Russian Federation towards the Weapons and Materials of Mass Destruction (WMD) Proliferation and Terrorism" showed that 69% of Russians believed Russia should participate more actively in international cooperation with the G8 countries in the field of biosafety and biosecurity, to prevent terrorist acts using biological weapons and fight against infectious diseases. To cope with the bio threat is a demanding challenge; thus it should receive more, not less, attention and support. The Global Partnership Principles to develop measures to account for and secure WMDs and related materials as well as to maintain effective border controls, export and transshipment controls are even more valid today than when they were adopted in Kananaskis, Canada in June 2002 [4].

The most difficult issue about bioterrorism is that its possibilities exist in water, land, food, air, and the human being itself. Biological agents that can be used for bioterroristic attacks are readily available, are relatively inexpensive to produce, store, and transport from one country to another. They can be toxic, transmissible and lethal; some have a long period of incubation. They are typically found in nature, but could be changed to increase their ability to cause disease, make them resistant to current medicines, or to increase their ability to be spread into the environment. Many items involved in biotechnology are dual use, thus they are difficult to ban. We have to recognize that the physical security of biological agents is very poor in a number of facilities. Insufficient border control makes possible illicit trafficking of dangerous materials, or weapons of mass destruction in the certain regions by terrorist groups [5].

Center of Disease Control (USA) gives the categories of bioterrorism agents/diseases, which are based on parameters such as lethality, toxicity, morbidity and mortality. For assessment of public health threat of each biological agent this classification is very important. According to it, category A is represented by high priority agents that "include organisms that pose a risk to national security because they can be easily disseminated or transmitted from person to person; result in high mortality rates and have the potential for major public health impact; might cause public panic and social disruption; and require special action for public health preparedness." Examples are Anthrax (*bacillus anthracis*), Botulism (*Clostridium botulinum toxin*), Plague (*Yersinia pestis*), Smallpox (*variola major*), Tularemia (*Francisella tularensis*) and Viral hemorrhagic fevers (filoviruses [e.g. Ebola, Marburg] and arenaviruses [e.g. Lassa, Machupo]).

Category B agents are those that "are moderately easy to disseminate; result in moderate morbidity rates and low mortality rates; and require specific enhancements of CDC's diagnostic capacity and enhanced disease surveillance." In this group we find Brucellosis (*Brucella* species); Epsilon toxin of *Clostridium perfringens*; food safety threats (e.g. *Salmonella* species, *Escherichia coli* 0157:H7, *Shigella);* Glanders *(Burkholderia mallei);* Melioidosis (*Burkholderia pseudomallei);* Psittacosis (*Chlamydia psittaci*); Q fever (*Coxiella burnetii*);

Ricin toxin from *Ricinus communis* (castor beans); Staphylococcal enterotoxin B; Typhus fever (*Rickettsia prowazekii*); Viral encephalitis; water safety threats (e.g. *Vibrio cholerae, Cryptosporidium parvum*).

Category C includes "emerging pathogens that could be engineered for mass dissemination in the future because of availability; ease of production and dissemination; and potential for high morbidity and mortality rates and major health impact" like Nipah virus and hanta virus [6].

2.2 Threat of Bioterrorism for Food Safety and Population Health

Different targets of bioterrorism attacks in the farm-to-table food continuum include crops, livestock, food products in the processing and distribution chain, wholesale and retail facilities, storage facilities, transportation, food and agriculture research laboratories [7].

Examples of diseases which were in the center of epidemiological surveillance and public health control in the last years include severe acute respiratory syndrome (SARS), foot-and-mouth disease, mad cow disease, monkey pox, and avian influenza. All of these infections were perceived as serious threats for population health by epidemiologists and health care authorities [8]. All levels of health care system including national, regional and local ones have prepared for the prevention and control of the mentioned pathologies. Many international organizations focused their activities on recording the cases, early detection, treatment of emerging infections and research of effective drugs and vaccines. Any of the mentioned diseases can be brought intentionally or accidentally to the country and have severe medical, economic, and social consequences.

The food processing sector is generally described as the middle segment of the farm-to-table continuum – it extends from the time livestock and crops leave the farm for slaughter and processing until food products reach retail establishments and the consumer. Terrorists could use food products as a vehicle for introducing harmful chemical or biological agents into the food supply. Toxic chemicals or infectious agents if contaminate food production facilities present potential public health threat [9].

Trends in global food production, processing, distribution, and preparation present new challenges to food safety. Food grown in one country can now be transported and consumed halfway across the world. People demand a wider variety of foods than in the past; they want foods that are not in season and often eat away from home [10].

The integration and consolidation of agricultural and food industries and the globalization of the food trade are changing the patterns of food production and distribution. These conditions are creating an environment in which both known and new food-borne diseases can become prevalent. Food and feed are distributed over far greater distances than before, creating the conditions necessary for widespread outbreaks of food-borne illness. In a recent crisis, more than 1,500 farms in Europe received dioxin-contaminated feed from a single source over a 2-week period. Food produced from animals given this contaminated fodder found its way onto every continent within weeks. The effects of exposure to dioxin from this source on public health may become known only after years of investigation. The international spread of meat and bone-meal prepared from cattle affected by bovine spongiform encephalitis (BSE) needs no further description. The full economic consequences of such incidents and the anxiety raised among consumers are still being assessed [10].

Other factors account for the emergence of food safety as a public health issue. Increasing urbanization leads to greater requirements for transport, storage, and preparation of food. Increasing wealth, an urban lifestyle, and sometimes a lack of facilities mean that people eat much of their food away from home. In developing countries, food is often prepared by street vendors. In developed countries, up to 50% of the food budget may be spent on food prepared outside the home. All these changes lead to situations in which a single source of contamination can have widespread, even global consequences. Developing countries in particular are experiencing rapid changes in their health and social environments, and the strains on their limited resources are compounded by expanding urbanization, increasing dependence on stored foods, and insufficient access to safe water and facilities for safe food preparation. The globalization of the food trade offers many benefits to consumers, as it results in a wider variety of high-quality foods that are accessible, affordable and safe, meeting consumer demand. A diversity of foods in a balanced diet improves nutritional status and health. The global food trade provides opportunities for food-exporting countries to earn foreign exchange, which is indispensable for the economic development of many countries and for improving the standard of living of many people [10].

However, these changes also present new challenges to safe food production and distribution and have been shown to have widespread repercussions on health [10].

2.3 Agricultural Sector as the Possible Target of Bioterrorists

Targets of bioterrorism can be the human population, branches of economy, lack of food, increase of prices, unemployment, market sharing, competition for new segments of market. Reasons for the threat of bioterrorism for agriculture are the following: biological agents do not present direct threat for human subjects, are not easily identified, artificial pest infestation is masked by natural epiphytoty or epizooty, increase of scale of international trade, unification of agricultural production by growing similar by genotype sorts of plants, the majority of farms and fields are not protected from bioterrorists, and the planning of large-scale attacks is facilitated by long incubation periods. An agro-terrorist could easily go into hiding or leave the attacked facility and the country long before the appearance of dangerous symptoms in the fields of agriculture. A bioterror attack against an agricultural facility is not only a psychological and ecological attack; it also produces a long-term destabilization of a system of food security in an entire region, causing rapid price increases for food before the expression of infection/intoxication symptoms [11].

The use of genetic weapons meaning the creation of genetic constructions introduced to the genome of the plant or infectious agent can cause epiphytoty or result in toxicity of the crops. These modifications can be maintained in agrocenosis, create reservoirs, join the co-adapted complexes of endemic diseases agents. High toxic strains of Fusaria, Aspergillus, and Penicillium cross by cross can infect plants, animals, and humans. Synthetic microorganisms can be produced so that they can possess pathogenic characteristics of fungi, bacteria, viruses [12].

Bio-ecologic weapons can be based on the use of alive infectious disease agents. From the mid of the twentieth century there was increase of bio-variability of pests and harmful organisms in agriculture, basically, saprotrophic fungi quickly evolutioning to parasitism due to introduction of intensive technologies of selection high-productive sorts [13]. The biological weapon can be also created on the use of modificators of the vital activity of grain while storing and processing crops with imported chemicals. High vulnerability of agricultural production and food chain for bioterrorism is explained by large amount of fields with genetically homogenous sorts of cereals, large granaries, and storages of chemical pesticides [13, 14].

There are some conditions, which increase the chance of bioterrorist attacks in plant-growing production. For example, if there is information about the absence of the reliable protective methods against biological weapon in the target area and no possibility of their rapid development in the place of the potential attack, we classify the area risky for bioterrorism. When the used biological agent possesses high harmfulness and has wide range of response on extreme conditions of environment, it is more likely to be used with terroristic purposes. If the potential damage to the crops is high but to recognize the aggression fact in the short period of time is difficult, then terroristic attack is more likely to be undiscovered that becomes a strong point of terroristic or criminal groups [12].

A special concern is presented by induction of epiphytoty of diseases in grain cultures. The action can be prepared in advance by the preliminary evolution of the pathogen in the region of planned epiphytoty (Ex.: race T of the southern helminthosporiosis agent in the corn was discovered in the USA in 1965 as a very rare isolate, epiphytoty that destroyed selection hydrides of the corn happened in 1970–1971). Accumulation of the enough amount of inoculum can be responsible for the occurrence of epiphytoty at several generations in the fields. Use of the common for the region disease agents, races of which are selected by high virulence for the locally cultivated sorts, is a risk factor [12, 14].

High danger comes from non-endemic but prevalent in the region agents (non- or slightly pathogenic) with increasing harmfulness, particularly producing toxins (some kinds of Fusaria, Alternaria, Aspergillus, Penicillium). Epiphytoty of fusariosis of the grain cultures happens every 3 years. About half of the grain crops in the world are affected by toxinogenic fungi. Exposing agricultural products to the toxinogenic fungi and contamination them by dangerous mycotoxins can have very severe consequences for food safety. Fusariotoxin and Aphlatoxins can be used by terrorists for contamination of water and food [12, 13].

Toxinogenic mycoses are highly probable agents for bioterrorism, because they have cross-affect on plants, agricultural animals, and humans. Toxinogenic effect on Y Chromosome that can cause conformation diseases, at which the necessary for life of the organism proteins are transformed to toxic ones and the organism dies from auto-toxicosis. Vulnerable for bio-terroristic affect are those food chains, for which agricultural products are the basic ones. Particularly, it is dangerous if the food is grown in ecologically polluted areas: if pesticides and heavy metals are present in the soil, the grain or feed are contaminated by toxinogenic mycoses or contain heavy metals; bone-meal prepared from cattle is affected by pryones, viruses.

Analyses of the probable scenarios for bioterroristic attacks include the following:

- To assess the vulnerability of each basic culture, to determine its critical points
- To make assessment of the process of growing, gathering, storage of crops, determination of vulnerability at each stage
- To determine the nature of affecting agents
- To estimate risk of affecting the basic cultures
- To develop the reliable and quick methods to detect the agents
- To work out the measures of public control
- To develop the actions for the critical situations (to protect the affected fields and to introduce the ways of full phyto-sanitary control in the region) [12].

2.4 Prevention of Bioterrorism and Provision of Food Safety

According to the EU Directive on Hygiene for Foodstuffs (93/43/EC), the HACCP (Hazard Analysis Critical Control Points) system must be the basis for safety procedures for all foods.

All food processors that process, treat, pack, transport, distribute, or trade foodstuffs are legally bound to have a HACCP plan. The processors of herbs and spices should rely on the HACCP system and develop their sanitizing program [10, 12].

However, the food supply is quite difficult to protect for many reasons:

1. The food system encompasses many different industries.
2. A variety of potential bioterrorism and chemical agents could contaminate the food supply, and the possible scenarios for deliberate contamination events are essentially limitless.
3. The public health system is complex, and responsibilities for prevention and control may overlap or may fall in the gray area between authorities of different agencies.
4. To achieve food and agricultural bio-security, the activities are needed in the areas of prevention, detection, response [15].

Prevention includes education for food producers about which bioterrorism and chemical agents would likely contaminate food, where in the production process contamination would likely occur, and what food-processing steps can be taken to eliminate or inactivate potential agents and chemicals. Guidance on how to assess plant or company vulnerability against a bioterrorist attack and to reduce the likelihood of a bioterrorist attack must be developed and introduced to practice [15].

Detection means the availability of methods for identifying credible threats, rapid and secure communication systems for sharing information on unusual events within the industry, enhanced laboratory capacity, the development of a primer for clinicians (including signs, symptoms, laboratory diagnosis, and treatment) on potential high-impact food-borne bioterrorism agents (e.g., food-borne anthrax, botulism toxin, chemical agents) to aid in rapid recognition of outbreaks. Guidance on developing action plans for response, including information on which agencies to contact for which types of events [10, 15].

Response includes the methods to increase government/industry coordination for investigating food-borne outbreaks, including issues of improving product traceability: improved coordination between animal health, public health, law enforcement, and industry for responding to bioterrorism events; guidance on developing streamlined systems for risk management communication throughout a product supply chain from farm to table in the event of a real bioterrorism attack; "just-in-time training" that provides accurate information in a timely manner to key industry leaders, employees, public health officials, and consumers in response to current events involving new or re-emerging disease threats [15].

2.5 Perception of Risk Related to Food Safety in Russia

By the report of the Russian Academy of Agricultural Sciences (2005) from 1991 to 2005, the services of agrochemical institutions in the country decreased 50 times. The reduction of public control on agricultural production, increasing amount of private farming sector, coming of foreign farmers to the national agricultural sector leaves possibilities for illegal use of toxic chemicals in agriculture and introduction of intensification technologies to get higher crops. Large amount of imported food products including meat and grain from the most of the world countries makes necessity on proper well-organized and effective phyto-sanitary control. Experts believe that joining of Russia the World Trade Organization in the near future can make the sanitary control of imported food worse, or at least it will complicate the regulation of imports. Under WTO, health and safety requirements must be justifiable on the grounds of protecting public health and must be based on a sound, scientific risk assessment [11].

We made a study on perception of risk related to food safety in the Russian population. The purpose was to evaluate the attitudes of the community to food safety, to determine the concern of people about the quality of products in the food market and healthy nutrition. By the specially developed questionnaire we interviewed 800 people in Kursk and Lipetsk regions (Central part of Russia). There was used cluster sampling; participants had age range from 18 to 85, response rate was 92%;

research was conducted twice – in the years 2006 and 2009. The results were analyzed by SPSS (version 13.0) for Microsoft Windows.

The major items of food security identified by the general public were addition of chemicals to food (74%), introduction of contaminants into the food chain from industrial pollution of the environment (52%), microbiological hazards in food (27%), new technologies like genetic engineering or irradiation of food (21%). According to our findings, bioterrorism is not perceived as high risk by the Russian population, but there is a trend of increasing awareness about it from 12% in 2006 to 19% in 2009 ($p<0.05$). In general people had more concern about the safety of imported products compared to the food produced in the country (81% vs. 32%, $p<0.005$). The highest trust was shown to the local producers of food (76%). It might be related to the well-known reputation of the regional producers and available information about them in mass media. It was found out that about one third of the ordinary consumers preferred the well-known brands and labels when they selected the food products, which were advertized on TV and radio. However, the major factors determining the choice of the product were the taste preferences and previous experience of tasting the food (65%), as well as health considerations (24%) and low cost (21%). People mostly did not recognize that they followed the advertisement while they selected food products.

Every third person in the study experienced consuming low quality food product in the last year and every fifth person had such practice in the last 3 years. People having negative experience (like food poisoning) become more concerned about food safety in the future, pay more attention to the reputation of the producer, expiration date, quality of package, become oriented at middle or high cost products.

When we interviewed the general population we found that the highest concern about food safety was in females compare to males (88% vs. 34%), older people above age of 60 compare to the younger ones below age of 60 (65% vs. 37%), more educated people with number of years of studying above 12 (74% vs. 46%); mothers of children under age of 12 (65%); all the differences were statistically significant ($p<0.05$).

The higher trust of consumers was found to be given to the food products sold in supermarkets and large stores compare to small shops and markets (65% vs. 35%). The particularly high awareness of people was registered about meat and meat products (sausage, ham, pate), fruits, vegetables, and canned food. The preference of the public is to buy the meat, fruits, and vegetables directly from the local private agricultural producers.

We also found that for 3 years of our study the concern of people about food safety has increased, people got more informed about the rights of consumers and the law regulating food safety, more often people started to claim their rights and defend them. We were glad to find out that people read more carefully the labels on food products paying attention to the ingredients of the product, nutrition value, presence of preservatives and GMO.

References

1. Centers for Disease Control and Prevention (2008) Bioterrorism overview. http://www.bt.cdc.gov/bioterrorism/overview.asp. Retrieved 22 May 2009
2. Frerichs RL et al (2004) Historical precedence and technical requirements of biological weapons use: a threat assessment. Sandia National Laboratories, Albuquerque (May 2004), p 11
3. Wein LM, Liu Y (2005) Analyzing a bioterror attack on the food supply: the case of botulinum toxin in milk. Proc Natl Acad Sci 102(28):9984–9989
4. US Department of State (2002) The G8 Global Partnership Against Weapons and Materials of Mass Destruction, Kananaskis, Canada, June 2002. http://www.nuclearfiles.org/menu/key-issues/nuclear-weapons/issues/terrorism/g8-global-partnership_Art-02-gp_stat-en_fco.gov.uk.pdf
5. Einhorn RJ, Flournoy MA (2004) Protecting against the spread of nuclear, biological and chemical weapons: an action agenda for the global partnership, vol 1. CSIS, Washington, DC, pp 27–28
6. Centers for Disease Control and Prevention (2004) Bioterrorism agents/diseases, emergency preparedness and eesponse. http://www.bt.cdc.gov/agent/agentlist-category.asp. Last modified 19 Nov 2004
7. Mackby J (2006) Strategic study on bioterrorism (Russian – European Union Group Study). CSIS, Washington, DC
8. Evans RG, Crutcher JM, Shadel B, Clements B, Bronze MS (2002) Terrorism from a public health perspective. Am J Med Sci 323(6):291–298
9. U.S. General Accounting Office (2003) Food-processing security: voluntary efforts are under way, but federal agencies cannot fully assess their implementation. U.S. General Accounting Office, GAO-03-342 Washington, DC, 14 Feb 2003
10. World Health Organization (2002) WHO Global strategy for food safety: safer food for better health. WHO, Geneva
11. Dyckman LJ (2003) Bioterrorism – a threat to agriculture and the food supply. Testimony before the committee on governmental affairs. US Senate/US General Accounting Office, Washington, DC, 2003
12. Monastirskiy OA (2005) Globalization of agricultural production and food safety in Russia/Food Safety in Russia. Materials of the Russian conference, 9 Dec 2004, Moscow State Academy of Veterinary Medicine and Biotechnology, named after KI Skryabin, Moscow, Edit House NP, 2005, pp 93–102. http://www.uvao.ru/okno/Prod_bezop.pdf
13. Dudley JP, Woodford MH (2002) Bioweapons, bioterrorism and biodiversity: potential impacts of biological weapons attacks on agricultural and biological diversity. Rev Sci Tech 21(1):125–137
14. Trudil D, Tartal J, Trudil C (2002) Experience received from the cases of bioterrorism in year 2001. Special symposium on problems of biosafety and bioterrorism (in Russian), St.-Petersburg, 28–29 June 2002. Chem Biol Safety N4–5
15. Center for Infectious Disease Research & Policy (2001) Bioterrorism and food safety: developing an effective national response. Center for Infectious Disease Research & Policy, University of Minnesota, Minneapolis, 12 Nov 2001

Chapter 3
Food Safety: The Global Problem as a Challenge for Future Initiatives and Activities

Radomir Radovanovic

Abstract Food is output of numerous and different activities inside very long and complexive "food/nutrition chain" and crucially relates to three important aspects. This aspects, being safety, quality and security, represent a basic right of every human. Namely, last few decades mankind has several very important global problems and among them, certainly, is one of specified aspects – FOOD SAFETY. Why? Because foodborne, as well as waterborne diseases, are major cause of illness and death all around the World. In fact, they are the leading causes of health problems and deaths in developing countries, killing approximately more than three million people annually, most of whom are children. However, food safety problems and diseases are not limited to one geographic region, political or social part of the world, but in different forms and shapes remain overall present. More than 9,000 deaths in the USA, due to food- and waterborne diseases, are certainly a strong indicator that the food safety is an issue in developed part of the world, as well. Thus, it is not surprising that in highly industrialized countries, such as USA, Australia and EU members for example, percentage of people suffering from different issue of unsafe food has been reported to be up to 30% (WHO 2003).

Having in a mind recent trends in global food production, processing, storage, distribution and retail, our state is that global problems with food safety is great challenge for the future initiatives, activities and strengthen comprehensive measures in all parts of "food/nutrition chain". Also, this will be a right reaction and clear reply on more and more stronger market and customer's demands – all in order to improve global food safety situation. So, the scope of this paper is to present and discuss some available data (*facts*) according to actual food safety situation (*problems*), as well as to suggest some initiatives, activities and measures that possibly can ensure a safer global food supply – safeguarding consumer health worldwide.

Keywords Food safety • Global problems • Trade • Regulations

R. Radovanovic (✉)
Department of Food Safety and Quality Management, Faculty of Agriculture, University of Belgrade, 11080 Belgrade – Zemun, Nemanjina 6, Belgrade 11080, Serbia
e-mail: laler@agrif.bg.ac.rs

M. Hefnawy (ed.), *Advances in Food Protection*, NATO Science for Peace and Security Series A: Chemistry and Biology 1, DOI 10.1007/978-94-007-1100-6_3,

3.1 Introduction

Within the last 20 years, we are the witnesses of serious global problems that the contemporary World is facing. Outstanding environmental pollution, drastic climate changes, hunger, and insufficiently fatting up and numerous incidents dealing with food safety are only a part of a long list of problems that seriously bring into danger the health and life of a contemporary man, as well as the existence of our Planet. The cause of the existing situation in the majority of cases is as per our attitude , the intensive development of technology, unsentimental run for profit, and hard to explain neglecting attitude towards obvious existing consequences, as well as to those that can be anticipated and expected in the near future. Obvious problems due to insufficiently and bad fatting up, numerous incidents caused by contaminated food and water, on one side, as well as the problems due to the wrong and immoderate nutrition, on the other, then quick living and working rhythm, a lot of stress situations etc. as a consequence of a large number of different diseases that seriously influence the health and life of a contemporary man.

The disturbingly low level of development, serious economical problems, then hunger and insufficient and/or bad nutrition, disturbing number of epidemics caused by contaminated food and water, as well as high mortality in certain World regions, many times have been the topics as well as terms of agreement among prominent representatives of the developed countries. The intention was, at least by declarative tone all the same: may adequate activities and before all, financial and expert support to help the most imperiled, and the existing unkempt status "food chain/nutrition". The role of the prominent functions is especially significant within administration institutions – both on international and national, local level. Their main task is to build a foundation, meaning to define optimally the policy and strategy in the area of food production, people's nutrition and health, to work on bringing proper legislative, but at the same time, to work on the development of the mechanism for efficient and effective application of rules in practice etc. Special role and responsibility are on experts in primer production, in the procedure of elaboration, modification, distribution and food traffic. A significant part of obligations and responsibilities are on all those whose products or activities give support to primary and/or industrial food industry. Last but not the least position, obligation, responsibilities are on scientific institutions and individuals that are devoted to research work in adequate fields of science of food and nutrition. And they, upon the results of research work, whole activities, reputation and authority must give a significant contribution to the improvement of production process, safety in supplying, security and quality of different nourishment products.

Accepting these obligations and responsibilities, we have decided to point out within this paper the actual and a part of the problem, but at the same time we have the courage to undertake certain initiatives, to point out some of possible activities, and to suggest certain solutions with an aim to improve the food safety. How much and will (?) those things and given suggestions be good and effective? The others will appreciate and the time that comes. Out of question is only – good will.

3.2 Problems and Consequences in Relation to Food Safety

Diseases caused by unsafe food and water (*foodborne/waterborne diseases – FBD/WBD*) are a serious problem of the modern World – existing not only in undeveloped or developing countries, but in the most developed industrial countries. More and more around the World a large number of FBD/WBD is appearing diseases, causing different unpleasant and human suffering: also serious economical consequences are present as well. The result of the systematic monitoring by WHO show that in the last decades, almost in continuity, health of more than one third of human population is more or less seriously brought into danger by biological contaminated food and water, and it is disturbing that a big number of cases end by death. More frequently the substances whose safety is not on satisfactory (and requested) level are registered, due to contamination of various chemical dangers, independent of the question of different residues, food additives or contaminants that originate from the environment. And while in the case of biological, specially bacteriological infections, the incubation period is relatively short (48 h to 7 days) and consequences appear mostly quickly, the negative cumulative influence of chemical contamination is postponed, but the human health consequences, almost by rule are very serious. The physical dangers that can seriously influence the human safety, although the consequences are primarily shown in various types of injuries. Moreover, the experiences up to now show that the cases of biological and physical food contamination are more frequently registered in less developed areas, undeveloped and developing countries, and at the same time, the examples of chemical contamination are more frequent in developed countries.

We are pointing out, that in the largest number of registered incidents, the causes of food contamination are established, in other words, causes for human diseases. In Europe the most frequent are biological dangers (about 97%) in other words bacteriological contaminants (about 92%). While the influence of chemical contaminants is significantly smaller (about 3%) – [1], in the USA the situation is similar. Although it has to be pointed out that during last 4 years (2000–2005) the number of incidents caused by bacterium is reduced for about 17% and the number of registered diseases is almost half ended [2]. But the high influence of incidents is disturbing, in which sources or caused by false food are not known. More over because these situations are in major cases present in the most developed countries of the World, meaning in the USA and Europe. Meaning in the score, number of the influence of unknown etiology incidents in the USA is modestly declining from 63% – 2002, to 58% – 2005, although the influence of cases of different diseases is in modest rise and it was from 35.5% – 2002, to 46.6% – 2005 [1]. Moreover, the figures that are related to 40 European countries, including the EC members, show that the cause of about 30% of incidents caused by false food in the monitoring period (1993–1998), was not known [1]. The actual situation is in the best way presented by the figures of the WHO, brought out in June 2005, during the International Food Safety Conference in Geneva [3, 4]. Due to those quotations, only because of diarrhea, about **1.5%** billion of people all around the world become sick, which is the most frequent death cause for the 5 years old children, and the young ones in the non developed countries – even about three millions per year. Apart of this, more than 1.5 millions of children, younger than 5 years become sick every year due to consuming water and food infected by pathogen bacteria, different viruses and parasites. How much is the situation alarming is shown in the best way by the facts by the WHO [5], according to which, in only 3 years, the children mortality caused by diarrhea for almost a million cases (2.1 million in 2002).

In some of our previous works, we have quoted numerous facts that point out that high developed countries are not spared of serious consequences caused by food contamination [6–9]. This time as well we point out the facts by the US Centre for Disease Control – USCDC, upon which, in this the World's most developed country, annually an average of 75 millions cases are registered of different acute diseases, caused by contaminated food, out of which, 325,000 are hospitalized, including about 5,000 dead. How serious the problem is?, is seen out of a relatively reliable assumption, that on every registered case, for example salmonellosis (*influenced by a pathogen bacteria Salmonella enteritidis*) there are 30–100 cases of this disease that are not even registered or reported. Due to the fact of the Center for Supervision and Informing about diseases caused by false food **(CDSD)** in England and Wales, annually about 1.4 million of cases are registered, out of which even 27.5% of patients have been hospitalized [10]. ***Frightening are the facts of the National Australian Bureau of Statistics-ABS***, that in this country on annual level, about two millions of cases of diseases are registered, caused by variously contaminated food [11]. More over, only salmonellosis (*as per 100,000 inhabitants*) are growing, and in about 10 years, was almost tripled: from **14** (1992) and **37** in (2001), to **40** (2003). So, unavoidably we come to a question: what is the situation and what are the consequences in less developed poor countries, if in the USA, where a firm legislation exists, and the monitoring and the sanctions are precisely defined, and are literally conveyed, annually close to 14 million cases diseases caused by food contaminated by different pathogen microorganisms?! Out of this number, about 60,000 are hospitalized, though even 1,800 comes to a tragic end [12–14]. More accurate information about the actual situation, consequences and costs in connection to food contamination can be found in WHO Annual Reports [3–5] and in the text by the author of this paper [8, 15–17]. Very useful are the details that embrace the number of registered cases on 100,000 inhabitants in some countries caused by different bacteria, or parasites or viruses [18]. Although understandable variations are visible among facts established for certain countries, commonly speculated is evident, that these facts only confirm the trend already registered in the World, due to which, the most common food bacterial contaminations come from the pathogen types as **Salmonella**, **Escherichia coli, and Campylobacter spp** [18]. In the majority of cases (*about 30%*), the cause of food bacterial contamination are errors or incidents in production processing [18]. The problems mostly appear because certain regimes of food health treatment are not fulfilled (*temperature, time*) independent to concrete process requests, or specific needed conditions for higher (*pasteurization, sterilization*) or lower temperatures (*cooling, freezing, warehousing*). The cause for bacterial contaminations is initial raw materials, non adequate hygienic technological equipment, tools and accessories, as well as the personal hygiene of the employees. Here, as well, is the troubling high influence of cases (>35%) where the origin of the bacterial contaminated food is not established [18]. Microorganisms, especially pathogenic bacteria, mostly threaten the food of animal origin. Here we think about eggs first, different meat kinds, milk and fish, as well as a complete range of different products that we get out of the process of treatment and processing of these important and highly valuable raw materials. Due to the official figures of FAO/WHO [19] exactly this group of products is mostly influenced by biological contamination, though different toxins which come as results of pathogen bacteric metabolites, not only that most frequently, but most seriously threaten the consumers' health and life. This is particularly present in situations when during the processing period the food is not properly thermally treated [18].

Generally speaking, the world food safety is influenced by major number of factors. As per our opinion, here, to be mentioned on the first place, serious problems in connection to food and water disposal (*so called safe supplying*), growth of international trade (agricultural and nourishing products), the enlargement of the number of changes in the population structure on the Earth, than the enlargement the range of traveling and migration movements, as well as the more and more demanding regulations on the international and/or regional level. Apart of the listed, it is not to be forgotten the influence of new technologies in primary plant and cattle production, significant new trends in the treatment processing and food packing, changes in food handling and way of preparing food, the growing bacteria resistance etc. Within this paper, we shall comment some of the more important questions and causes of global problem in connection with food safety.

3.3 Causes for Arising Problems in Connection with Food Safety

3.3.1 Problems in Supplying Food and Water

According to the official data of the United nations (UN) for 1990, more than 1.2 billion people or about 28% of the developing countries population, live in the conditions of extreme poverty, meaning survival with daily income less than 1 US $ [20]. The truth is, in 2002 this part was reduced for about 20%, but due to the FAO evidence (*Food and Agriculture Organization*) one of five people in undeveloped and developing countries still has permanent and serious problems with food missing. Because of that, over 840 million of people on the Planet, suffer from hunger (*to some sources 1.2 billion*) and even about 30% of World populations suffer the consequences of bad and insufficient nourishing, unfitness or by using unsafe (*contaminated*) food and water. All that, influences the high mortality, as well as numerous different health problems. Even about 55% of total 12 millions of children die annually because of the consequences of some of the listed food problems. Besides, about two million of people suffers from insufficient iron level and anemia, out of which 52% are pregnant, and 39% are children under 5 years of age; the missing of iodine brings into danger the health of the 740 millions of people, and the lack of A vitamin influences even 100–140 millions of children in the World etc. [21].

Within the agreement to overcome this situation in the most efficient way, at the World Food Summit, 1966, the goal was created, that up to 2015 hunger/unfitness in the World should be haltered, meaning that up to that time, the number of brought to danger people should be systematically reduced for 22 millions per year. Due to up to date figures-this task is difficult to be achieved, as the reduction on annual basis now comes to only six millions or 27% out of planned. Besides, threatening is the information about the application of new standards of the WHO relating to child growth standards in the World, as the new methodology enables to get more sufficient data on global unfitness, especially with children. Therefore due to the new methodology of information treatment, 388 national data basis from 139 countries [21] proved that in 2005 (twenty-first century) in undeveloped and developing countries, even 32% or 178 million of children over 5 years old, are suffering unfitness due to insufficient food supply. Out of that, more than 40% lives (*more precisely survives*) in the regions of America and West Pacific. Out of 39 mostly hit countries in which more than 40% of the population is threatened, 22 are from Africa, 7 from South-East Asia, 4 from East Mediterranean and 4 from West Pacific, 1 from Europe and 1 from America. Out of 35 countries, in which 20% or less population is brought to danger, 13 is from the region of America, 11 from Europe, 6 from the area of East Mediterranean, 3 from West Pacific and 2 from South East Asia.

Data of the relationship between body mass and height were used for statement of account among the children, show that the acute unfitness is a serious cause for less, meaning that due to the level of unfitness, the mortality among children, who are the most fragile part of the world population, can be foreseen. So by applying of the methodology of the new WHO standards, forecasted that by this way caused losses can come up to 55 millions of children over the age of 5. Among them, it is forecasted that the largest number brought to danger (*about 29 millions*) live in the South East and Central Asia, and that in other regions the same situation is shared by 19 millions of children. Even more, a significant number of these children die even earlier, meaning before the age of 5. Generally, compared to the previously established data, it is obvious that the threat of all population groups is 1.5 to 2.5 bigger from the period the methodology of new WHO standards is applied. Specially threatening is the information that the influence of the brought do danger is larger, and the fatal outcome more frequent at small children and the young ones. So, the problems are significantly larger. The actual situation is very serious, while, numerous activities and steps undertaken by FAO and WHO, especially within the World Food Programme, obviously are not efficient enough, and they do not give the expected results. As an example, we can see, that by the mentioned programme only in 2000, 83 million of people were comprehended out of 83 countries out of that about 12 million of school population in 54 countries, but the registered consequences of hunger and unfitness in the World are not significantly reduced. So, the direct help in food, or precisely speaking, "actions" in food delivery, give certain partial results, but it is no doubt that, this way of help, for a long period of years, does not essentially overcome the global problem of hunger and unfitting of a significant part of the population [22].

Far better results could be achieved, if, due to our attitude activities and concrete movements would be directed to long term and serious overcoming of them, in essence, key problems. Here, we primarily think of the necessity of solving problems connected to regular supplying with clean drinking water (*at disposal*), watering the ground and crops (*optimum and rational water utilization*) and population education unthreatened areas (with an aim to instruct them to help themselves on long term basis). Why? Because one of the more serious global problems in the greater water shortage, and due to the FAO data, in the World, for the agricultural needs, about 70% of total water on disposal is spent while, in the developing countries, this usage is higher, coming from 85% to 95% [23]. The seriousness of problem is pointed out by data that show that watering of cultivable soil in the world is saline and inefficient, and at the same time very uneven among certain regions, and it brings to big differences in rational usage of available water qualities. For example, in Asia only 42% of cultivable soil is watered, on the Near East and North Africa 31% in Latin American and the Caribbean's about 14% while, within the area below Sahara only 4%. As watering can extend the contribution of crops even for 400%, in the coming 30 years, it is necessary to enable at least 70% of the total production of cereals to be from the surface that are watered. In connection with that, FAO is planning that until 2030, watering of agricultural soil and crops in threatened areas of the World should be enlarged for about 27%, but the improvement of efficiency of the irrigation systems, meaning their rational usage, will enable far more effective production including the larger water consumption of only 12%. For the relationship of listed and other FAO and WHO programs, meaning that hunger, unfitness, diseases, and big mortality (*specially children*) in the World would be significantly reduced, according to our opinion it is necessary among other, that eight the richest countries (Group 68 members) and other devoted countries give a serious help. The help should not be "merely" creating funds, "actions" of charity and partial solving of crises situations – as it is very often in the Sudan, Afghanistan, Malawi, Chad, Ethiopia, Mozambique and many other countries. It means that all further activities should be well planned, top level synchronized and coordinated, including full cooperation with FAO and WHO and with maximum engagement and help of their experts.

Previous remarks and comments are made essentially for the reason that about the food and water safety cannot be serious talks, if it is not seen, the existing of such great problem is fulfilling the basic condition for life and civilization right of man, and that is ferret to be used, meaning being sure in water supplies and essential food. More to say, listed problems in supplying are tightly bound with general hygiene life conditions, large epidemic as well as more or less expected consequences of the incidents caused by polluted (*contaminated*) food and water.

3.3.2 International Trade of Agricultural-Nutritional Articles

The process owners (stakeholders), experts and researches who performed their business activities in different fields of "food chain/nutrition" know that one of the leading causes of food contamination and large number of incidents, which are as a rule followed by adequate consequences, is the intensive international exchange of agricultural/nutritional goods. This is confirmed by the fact that the range and value of the international traffic of food, from year to year is in permanent progress. Meaning due to the recent fact by FAO [24] in the 5 years period (2000–2004) the total average value of food traffic in the World was about 984 billions of US $ (*import about 504.8 and export about 479.3 million US $*) in the quoted total value the major influence is within the group that embraces traffic of cattle and animal origin products–an average of **672** million US $ or about 68% [24].

From the point of view of food safety, special significance has the growth of international exchange of nutritional which can be seen in all regions of the World; as in the fields to which the highly developed countries belong, the leading food producers (*for example Europe, North America, Australia*) and in the fields in which undeveloped countries dominate together with the developing countries (*Africa, Asia, Central and South America*). It means that outline average value of foreign trade exchange of food (*import-export*) in the 5 years lasting period (*2000–2004*) for the countries in Africa was about 38.8 million US $ in Asia-210, Oceania-30.5 (*the share of Australia is about 24 million US $*) for USA-108 and Canada-31, and then for the countries of Central America 28.3 and South America about 55 million US $. The largest value in the international exchange of food is enabled by European countries – in average about 476 million US $ per year, out of which only 25 countries EC members – not counting the internal exchange within EC-enable annual exchange of food with a value of a bit more than 30 million US $ [24].

Due to the same source of information in the previous 5 years, Serbia has enabled a significant foreign trade food exchange, the average value, on annual basis is about **1.1 billion US $** (*export about 636 million and import about 474 million US $*).

3.3.3 International Regulations Related to Food Safety

Serious global problems related to safety and permanent increase in food traffic in the World, have traced many activities in the attempt to establish as efficiently as possible international regulations, whose consequently application should mean a significant improvement, the hygienic-toxicological and every other correctness of food in traffic, and by that to enable high level of consumer's health protection (*safety*). Here, to be mentioned at first is the significance of two International Agreements: on the application of Sanitary and Phytosanitary measures – SPS from 1995 as well as an overcoming of *Technical Barriers to Trade Agreement – TBT* from 1979. Of extremely high value is the promotion of the request and larger application of the up-to-dated concept of the risk analysis and the critical control point (HACCP) but it requires the documents CAC/RCP 1-1969, Rev.4-2003 (*Recommended International code of practice-general principles of food hygiene*) whose consequent application means the basis and condition of the implementation of HACCP concept [25]. At the end, very important are the recent documents of the European Commission that make the direct normative support to the food safety, as the Food Law of EC (*General Food Law, 178/2002*) as well as the directions of EC 852/2004 – with an obligation of application within the country member from 01.01.2006.

We have pointed out that one of the most important documents in the area of food safety on the international level so called **SPS Agreement** [26] and is referring to the animal and plant health protection as well as to the safety of nutritional goods. The Agreement text was accepted and published within the Final Act so called Uruguay Round of the international trade negotiations, signed in Maraquesh (Morocco) [27] and in the countries members of the World Trade Organization it came to life January the 1st 1995. In it, besides the SPS agreement, there are adopted amendments on GATT agreement (General Agreement of Traffic and Trade) from 1947 and within the Uruguay round of negotiations, a decision was made, to establish WTO as the supreme international body which has to deal with question of World trade. So, WTO with 132 member countries in the moment of establishing, became practically the inheritor of GATT, meaning that it undertook earlier functions of this organization, with a significant influence, as well to rule defining, as to its complete streams of international trade of agricultural-nutritional goods.

Within the SPS agreement, sanitary and phytosanitary measures are defined as any other measure that: protect man and animals from risk influenced by the additive consumption, presence of contaminants, toxins or organisms that might cause diseases and which are in their food; protect people from diseases transmitted by plants and animals; protect animals/plants of mischief doer, diseases or organisms that might cause the disease; prevents or restricts all means of damage caused by entering/developing/ spreading the mischief doer. Measures of the protection of the environment, the question of animal benefit in a wide meaning, as well as situations relating to the consumers' interest protection are not subject to SPS Agreement. This topic is dwelt by and it is clearly defined by other WTO documents among which we specially emphasize article XX GATT Agreement from 1994. This Agreement allows countries to define specific conditions (requests) and undertake corresponding measures for imported goods "necessary for protection" the human health and life, as well as animals and plant health and life, even when these requirements are more strict compared to those that are related and applied to domestic goods and international trade. This article has as per our opinion, left space for economic protectionism in international trade. That is, so to say, these situations might appear under the circumstances, when the governments under the influence pressure of local facts have to go beyond the protection of its own population, applying too restrictive measures, what is mainly in the function of protecting the interests of local producers against foreign trade. We particularly stress this, having in mind that SPS Agreement's main function is to promote the supreme of each state to achieve the optimum protection level of its population, but to enable at the same time that this right should not be misused, not having as a consequence building of not needed restrictions in the international trade [26]. Specially because developed countries with highly arranged systems of food control, by import have the possibility to undertake restrictive protective measures towards the goods origined from developing countries, for the only possibility to approach the world market. Frankly speaking, the national sanitary and phytosanitary measures and standards in the undeveloped and developing countries are very often below the requests that are defined at an international level, and that's why the SPS Agreement application for the developing countries has been postponed until 1997. and for undeveloped countries until 2000. It means, that these countries have been "abolition" for some time from the scientific confirmation and adequate explanation of own sanitary and phytosanitary measures – but only until the proved time limits. Many of them, faced with the lack of material and human potentials, or because of very practical reasons, have accepted the standards created by professional international associations (CAC, OIE, IPPC) as their national rules. That is how the undeveloped and developing countries enabled requested rules which enables them placing nutritional products on the market of other countries, independent of their own economic power or less sophisticated technology used for getting these products-but only to reaching the equal level of safety as in the country of the importer. Having in mind numerous difficulties that these countries are faced with, especially the significance that placing on the international market have their nutrition products, SPS Agreement gives the opportunity to the governments of these countries, not to apply these international standards completely, as the accepted risk level can be achieved in different ways, but only under the condition that it can be confirmed that applied measures the governmental level of the consumers' health protection. Therefore, the governments (states) are encouraged to accept that package of measures in these situations which enable them the same protection level, introduces less restriction or prohibitions in trade. Nevertheless, if national requests some of developed countries, as a consequence have a significant restriction in trade, WTO can undertake appropriate protecting measures, meaning to expect from the member country scientifically based evidence which has to be clearly demonstrated, why in that country conditions, the applied measures are not able to provide expected level of consumer's health protection.

Another very important agreement in relation with international trade is TBT Agreement (*Technical Barriers to Trade Agreement*) from 1979. It developed out of so called Tokyo round of trade negotiations in the period between 1974 and 1979, and it is known as a "***Standard Code***". TBT Agreement embraces numerous technical requests, standards and procedures, which, in the broadest way apply to marketing of various goods. As concerning nutritional products, TBT Agreement is a signed technical regulation that embraces different measures of consumers' protection – except for sanitary and phytosanitary. So by TBT Agreement, and its measures, the international standards regarding the type and quality of packing for nutritional products have been promoted, instruction for its declaration and marking, requests concerning nutritional abilities of food etc. [6, 28].

The function of SPS and TBT Agreements show itself, among others in promoting the rules and adequate practice in the preparation process acceptance and application of adequate regulation by the responsible national (governmental) bodies. As well, important role is in the optional harmonization of national regulations with international standards which are an activity result of authorized international associations and bodies. Our stand point is that the positive effect of SPS Agreement application in relation with enlarged safety of food and consumers' health protection, obvious in the majority of countries WTO members. This means that this agreement strongly supports the food safety high lightening the systematic application of scientific knowledge, and at the same time, in a certain way, makes more narrow the area for protectionism and decision lacking scientific support. At the same time, as a result of government's transparent decisions and procedures, the consumers have more accessibility to information concerning food safety, human animal and plant health protection. At the end by eliminating needless trade barriers, direct consumers are enabled to have a greater choice of nutritional stuff, safe for health-all as a result of growing international food traffic and "healthy" competition among producers [26].

Numerous global problems that are appearing today, in relation to insufficiently food correctness and threatening the consumers' health, especially in foreign trade turnover, the consequences are among others inadequate harmonization of rules in this area on international level. A significant contribution is overcoming such a situation, had, and they also have today, two important international associations, established immediately after the Second World War. The first is FAO (*Food and Agriculture Organization*) established in 1945. with authorizations and responsibilities in the field of agriculture, food and nutrition, while the other, WHO (*World Health Organization*) formed in 1948 with authorization and responsibilities which relates to human health protection and producing adequate regulative, embracing as well, regulations concerning food. During the WHO Assembly (*16 May 1963*) a specific FAO/WHO Programme was accepted (*Codex Alimentarius International Food* Standards *Programme*) and a specialized body – *Codex Alimentarius* ***Commission (CAC)*** was formed.

The essence of the mission and concrete activities of this international body (CAC) as it is pointed out in its establishing act, in the first article of its Statute, are consumers' health protection and providing equal honest (fair) conditions in international food trade. How big its significance is, is best seen within the UN General Assembly resolution from 1985 (published in 1986) declared as "***Consumer's Protection Direction***". So to say, in this document food is recognized and defined as one of the priorities which are of essential significance for human health, and in that sense, specially pointed out that when food is in question, the only reference is – ***Codex Alimentarius Commission***. The aim of this body among the others is to expose all significant questions that relate to consumers' health protection, and are in any way in connection with food, to researching projects and studies, and upon the results of these researches make a unique international regulation on food. That is how CAC provides adequate standard support (base) and to the international community pays attention to an extremely important area of food correctness and quality, meaning the question of its maximum safety when the consumers' health is in question.

In spite of numerous difficulties in the process of harmonization of regulations in food, which was specially influenced by difficulties how systems and law forms of 163 countries FAO and WHO members (*embracing about 97% of world population*), CAC by large engagement made a significant contribution that more countries, their national food standards, specially those bound to correctness and safety, coordinate with standards and acts proclaimed by CAC. Therefore, the food law ***Codex Alimentarius*** became an essential, global and referring base for food consumers, producers, and processors, national food control agencies, as well as for international trade of agricultural – nutritional products. Application of this regulation spread to all parts of the world and the majority of countries, and because of that, its particular contribution to human health protection in particular relationship in international food traffic cannot be measured. As well the existing regulation CAC means an unique opportunity for all the countries to join the international community in the defining process and the harmonization of food standards as well as working on its global implementation. The advantages of such a system of protection are clear by itself. Exactly because of that, standards of this regulation (law) became a stand point for comparing all national measures and regulations in connection to food production and traffic.

Codex Alimentarius Commission brings and adopts particular ***standards, recommendations and directions*** all according to the procedures regulated by the statute of this body. Meaning the activities in connection with defining of mentioned regulations can be split in two essential groups, according to the split of specialized committees (*Codex Committees*) which proceeds then to CAC commission to be accepted. So, there exists so called *horizontal* group within there are committees (*as well as appropriate standards*). That deal with questions of common significance; here nine committee achieve activities, among which is the ***committee for Food Hygiene***. In the other, so called ***Vertical Group,*** are committees (*and adequate standards*) that deal with the problems significant for a special group of nutritional stuff: here functions 16 (sixteen) committees among which are for example***, committee for meat, hygiene, committee for meat and poultry production etc***.

One of the numerous documents (there are over 200) accepted by now by the Codex Alimentarius Commission in ***Recommended International Code of practice – General Principles of Food Hygiene CAC/ RCP – 1 (1969), Rev. 4*** [25]; annex contains direction *"Hazard Analysis and Critical Control Points System and Guidelines for its Application"* – HACCP . By this document, HACCP Concept is officially promoted and recommended as a system which through whole *"food chain/nourishment"* has to achieve adequate level of safety concerning the safety food for human needs-from essential production to final consumers, meaning in the international traffic as well.

The influence of the mentioned ***Codex International Food Standards*** *Programme* on the international trade and food safety is extremely big. This program creates space for the leading world experts covering specific areas of food production and traffic to achieve, through discussions and opinion exchange, scientifically based agreement for certain questions of mutual interest. In the progressing global economy, achieving of satisfactory level of food safety is a mayor and very important task, which can be hardy achieved by a country itself, by individual and isolated activities- specially undeveloped and developing countries. Therefore, CAC solutions mean, in essence a wide agreement of scientific approach dealing with the regulation in the field of food safety, but a wide possibility of application of this regulation in practice [29]. However, it has to be pointed out, that there are still those that like ***Jiranthana, 1994*** [30] state possible problems in connection to previously mentioned standards and recommendations in the light of SPS and TBT agreement. Therefore, the doubt is thrown upon the possibility that subjects in international food trade proclaimed HACCP concept as a condition and obligatory – *specially in cases when the food exporting country can scientifically explain and prove that not using HACCP approach in production and inspection it can achieve equal level of food safety, as well as in the country that imports food, and which the application of HACCP concept is obligatory as per appropriate natural regulation.* Although theoretically observing, this scenario is possible, as insisting on the obligatory application HACCP concept would principally be breaking down not only SPS but TBT agreement in international trade as well. It has to be pointed out that such a situation, at least up to now in practice of international trade of nutritional stuff has not luckily happened yet.

Previously mentioned situation (*and the rising questions*) concerning the application of HACCP concept is not alone. It means, there is a dilemma concerning the existing of so called *"performance standards"* upon which fulfillment of each HACCP system is guided in the sense of projecting its efficiency, or expected efficiency (*results*) that have to be lowered to an acceptable level of the food safety risks. So, we come to a question if the HACCP instead of *performance standard*, becomes *processing standard* in the situations when it enters into the frames of corresponding regulation (*for ex. Rules, regulations etc.*)*?* The answer to this question depends on concrete solutions that will be taken from the governments of certain countries in the sense of projecting national regulations which make order in application of HACCP concept. As per our opinion it will (*at least by its mayor part*) stay as a standard of performance – of course under the condition that it will be accepted by the legislator only within 7 *(seven)* initial and widely accepted HACCP principles, proclaimed its application as obligatory, and without specifying and proclaiming numerous individual request in details. That is the case today in the majority of European countries (*with an exception of England, Scotland and Ireland*) including Serbia. Only when by law or appropriate HACCP regulation and recommendation precisely define and proclaimed for example technical – technological conditions for objects themselves and their environment, temperature and cooling time, conditions of technical treatment of each product (kind and type), then kind and way and frequency of sanitation within and out of the operation and similar, of course within the general philosophy of HACCP concept, this system is becoming a processing standard. On such a base, HACCP regulation incorporates an obligation of setting up of certain critical control points, their critical borders, promoting adequate procedures for monitoring, evolution, verification, conditions, way and dynamic (*frequency*) of its handling and so on. This situation is at least at the time of writing of this paper, recognized only in practice in the USA and Great Britain. Well, will the

HACCP concept be applied within the WTO country members as a ***standard of performances***, or as a ***processing standard*** it will be finally decided how will HACCP function as standard in international trade [31].

A part of mentioned, there exists a realistic possibility that in practice requests should be put, according to which it is necessary, in scientific transparent manner, come to a reciprocal comparison of *different standards of performances*. Such a task can be put in situations, when with an aim of removing possible trade barriers, with a wish to prove that to different HACCP systems – although established matching with 7 main principles but conducted as per different standards of performances – achieve the same level of food correctness, meaning consumers' health, protection, and safety. Out of up to now practice, at least one lesson is clearly understood: HACCP can function as an strikingly and effective system in international food trade only under the condition that the HACCP requests of so called preconditioned programmes (*good manufacturing practice – GMP and sanitation standard operational procedures – SSOP*) would be coordinated and joined at the international level. For this reason, because if possible differences in GMP and SSOP requests, might be equally "striking" and "effective" protectional barrier in food trade, as well as in the cases of own HACCP concept request [31].

At the end, with no intent to give a complete comment process of certification, we only want to point out that the place and role of certification of HACCP system from the **"third"** side, in the sense of official confirmation of achieving appropriate effects in relation to the food safety, on the international level has to be defined and considered in details. The international **"jury"** dealing with certificate directions is still at session, so the new extended solutions might have answers, primarily within the attitudes and agreements of the governments, professional associations and operations food manufacturing of certain countries. The author of this text has the opinion, that for some time in the future CAC will not (*we think it does not have*) directly define and undertake the arbitration role on all questions. The part of the problem will be solved as activities of accredited, especially *"aggressive"* certification bodies, which in all that we can see big business interest, especially from the moment of the official declaration of the international standard ISO 22000:2005. This typical managing standard, gives foundation to accredited and certificated bodies for handling corresponding activities. Of course, once more (*and who knows how many times up to now*) we repeat that here we put up *"only"* or at least three questions:

1. Will and in what extent will one side appreciate the certificate of any certificated body in a concrete export – import business arrangement, or it will act according to appropriate local regulation, or upon own choice of so called ***"other side"*** that has to do the working conditions checking (*PRP – GMP/SSOP*), the effectiveness and striking of HACCP system of a producer?
2. Will – as it is shown in practice in the cases of QMS (ISO 9001) and EMS (ISO 1400) consulting companies, certification and accreditation bodies be relieved of responsibilities in the case of appearance of serious incidents caused by the side of certificated producers?

3. Do the existing low systems recognize, meaning do they have instruments for regular orientation of action all already done, and the consequences of the offender within a criminal and offence or any other responsibility?

So, there are many questions with no precise answers yet. It points out that in the near future a highly professional and responsible work is inevitable in solving numerous existing dilemmas and exact problems. The only conditions that are not questioned here are great moral responsibility and serious financial and business consequences. All members of the "food chain/nutrition" might be affected by these conditions provided that they are responsible in the case of any, especially serious incidents in connection with the food safety and health risk to the consumers.

3.3.4 Population – Numbers and Structure

The total population on the Earth in 2000 was more than 6 billion, with a projection that in the following 50 years, it will increase about 32%. Meaning, in undeveloped and developing countries will increase 60% and in developed, as well as countries in transition (*including Serbia*), will increase less then 5%. Particularly disturbing fact according to which the growth is predicted, is that the most intensive population growth is in the regions which are the least developed, and which today have serious problems in supplying sufficient quantities of safe food and drinking water (*in Africa for 38%, in Asia for 41%, in Latin America for 52%*); only in Europe, in the researched period (2000–2005) forecasted in the declaration of total population for about 10%. Having in mind the dynamics and the regional spreading of the population growth on Earth, in the time in front of the human population, it will be necessary to undertake numerous, well concepted and synchronized activities – all with an aim to enlarge the producing volume, improve more significantly good producing and hygiene practice, meaning to enable much better disposability of safe food/water of good quality.

From the aspect of food safety, it is significant to stress obvious changes in the population structure in the World [32]. It means, that in the last 45 years (1960–2005) the number of urban population is in expansion for 2.2% and rural for 1.3% million. Due to the forecasting for the following 10 years period (2005–2015) the number of inhabitants in towns will enlarge for new 888 millions, as the rural population will enlarge for about 100 million. So, in the whole researched period of 55 years (1960–2014) the influence of rural population will increase for about 3 billion or 4 times, as the influence of urban population will grow for 1.4 billion or about 70% [32]. These facts are listed for at least two reasons. First, apart from the fact that the agricultural production is distinguished by more intensive and automatized production, number of people that directly take part in the manufacturing process of raw materials, for producing nutritional stuff is decreasing. And second, the number of incident and cases of diseases caused by unsafe food is more frequent in urban communities.

At the end, we want to point out that the age structure on the Earth is unfavorable. Meaning, in 2005 the share of inhabitants of 60 years of age and more was 10% of the total population, while this share is much larger in developed areas (*about 20%*) compared to undeveloped and developing countries (*about 8%.*) Frightening are the prognosis that until 2015 the share of this age group in the world will grow up till 12%. We are mentioning these facts just because older people (*consumers*) are more sensitive, their health is more frequently threatened and consequences of consuming unsafe food are harder to be settled and longer as well, specially compared to younger and middle aged.

3.3.5 Traveling and Migrational People Movement

Increased number of incidents caused by unsafe food and water is significantly influenced by more and more intensive touristic, business and in some other way motivated trips of the population, all around the Earth. Due to the facts of the World Tourism Organization during 2005, the total of *officially registered* international trips was larger than 800 million. The largest part was completed within touristic trips-402 million (*50%*) business trips influence to 125 millions (*16%*) as for 211 million cases (*26%*) the motives were very different (visiting relatives and friends, curing, recovering, attending religious gatherings, sport competitions, cultural events etc.). As per 8% of trips we do not have the reasons [33].

Apart of registered a significant number of international trips on annual level is not registered as they are organized and achieved on personal basis within individual and/or family traveling arrangements. We can almost surely say that the total trafficking of passengers is about one million per year. If we add organized yearly trips within the boarders of a country (*internal-traffic*) to this number, then we can forecast the range and intensity of a total human trafficking on annual basis in the World.

This information clearly shows that large people movements and staying in the environmental different to those people are coming from, and to which they are used, meaning larger exposure to different conditions and different risks they are used to, come to a smaller or larger instant threaten the passangers' health. Due to the WHO facts, the cause of the largest number of registered infections, gastrointestinal and other health problems of people during the trips is appearing just as a result of consuming unsafe food and water, sudden change of climate conditions (*such as temperature, humidity and the atmospheric pressure, than above see level*), specific insect kinds and microorganisms influence to increasing body sensuality and help to the easier contamination of passengers. The incidence is more frequently registered during the visit and stay in undeveloped and developing countries, meaning in the environment in which the average hygiene – sanitary living conditions, including manufacturing, local traffic and food preparation on certainly lower level compared to developed countries [33]. The problem is even bigger if we have in mind that just in these countries many touristic interesting destinations are, beautiful landscape, valuable sights of old civilizations, cultural monuments etc. (*Asia, Central and South America, Africa*) causing great interest to visit these countries.

And, as we have pointed out, just in those regions of the World, the major problems are registered in connection to supplying (availability) and food and water safety, so it is understandable not only the phenomenon but real trend as well of the increasing number of incidents caused by contaminated (unsafe) nutrition food.

The frequency of incident appearing, seriousness of diseases and consequences as well as the possibility of first aid and curing the ill is influenced by a large number of facts. For example, the smallest number of incidents caused by unsafe food is now registered during business trips. The reason is that visitors are targeting higher level hotel accommodations. The majority of time they spend in comfortable conditions of a business area, conference centres, etc. the meals are prepared applying higher quality and safety standards in well controlled conditions. On the contrary, the largest number of incidents are registered during individually organized trips, when, primarily for financial reasons and saving possibilities, previously not checked accommodations are booked, and water and meals are usually consumed "by the way" in smaller restaurants and fast food shops where the total level of hygiene is very often under suspicion, and sources for supplying of food stuff and safe control of food preparation – insufficient.

In the traffic structure, the largest number is realized by flights – about 45%, while the railway and traffic by water are used in 25%, meaning in about 8% of cases. The previous facts we list, before all, because the smallest number of incidents happens by consuming food and beverage on aircrafts (*less than 1%*). This is quite understandable, and it can be easily explained. Flight companies in the majority of cases supply food and beverage from their own production (*catering*) or, during their requests and under their trade mark the complete meals or part of them are prepared, within aimed production among specialized manufacturers. In both cases the production process, warehousing, distribution, delivery, keeping and serving the food are defined precisely, and systematic control of food safety is on a significant level. Big attention that is paid to food and beverage is quite understandable at least for two reasons. First, the destinations are often very distant (*few thousands of km.*) and the flight durations and length are sometimes very long (*more than 10 h*) so the consequences of a possible passengers contamination by unsafe food could be very serious. In flight conditions, the possibility of serious medical help of possibly threatened passengers is almost minimal. To highlight the significance of these questions we quote an information of the International Civil Aviation Organization – ICAO, that during 2005 it was registered more than two million flights in the World. The prognoses are that in a period 2006–2008 number of flight will rise as much as 6% per year.

Situations similar to the previous are characteristic for railway transport as well (25%) and by water (8.5%) as in this mean of transportation we deal with very organized and well controlled food and water supplies. This in particular applies to modern railway companies so called "quick rails" as well as to bigger river and transoceanic boats.

At last, within this part of exposure, we wish to mention another large group of people, who are forced to move frequently, locally or globally, thus changing life conditions. This is not a question of tourists or business people who are following different motives and willingly travel around the World. We think about a large number of unfortunate people who have to travel, frequently changing places to stay because of pure survival. We think of emigrants and refugees. As per WHO facts, international migration of people during the past 15 years has been increased from 120 million in 1999, to even 190 million in 2005 or, about 63%! The reasons for migrations are numerous and very different:

- economic (*searching for job and/or better income*),
- political (*dissidents and political emigrant*) the situations caused by wars, resulting in numerous cases of refugees,
- natural disasters (*loss of homes, positions, and properties*).

This problem is very complex and serious. Data show that in some countries immigrants make up to 20% of the total population (*way higher in some countries as Canada*). Within the topic of this paper, we stress that people are forced to migration movements. Especially refugees are very much exposed to risks of, contamination and numerous consequences caused by consuming unsafe food and water. This is specially seen in the countries facing great problems with their own population referring to enabling sufficient quantities of safe food and water supplies as Ethiopia/Eritrea, Sudan area below Sahara, parts of Central America, East Asia etc.

3.4 Safety Food Improvement of the System for the Development (Possible Solutions)

A principle widely accepted in the World, according to which, direct responsibility for the food correctness and the final consumers' safety is on all who by any mean take part in the "chain food/nutrition". This applies not only to the companies (*so called joined responsibility*) but to high functions representatives (*personal responsibility*). We specially stress that responsibility is not avoiding the institutions and/or individuals, and facts confirm it, from "the bask rows" meaning inspectors, inspections services, auditors certificated accreditional bodies, consultants and consulting agencies, authorized suppliers of specialized services etc. So, when an incident happens, and people's healths and lives are more or less threatened, or when the investigation and court procedure proves responsibility (i.e. following adequate sanctions). For example, only within the first 6 months of the 2005 upon court decision in USA, the elevation of stated sanctions to companies and individuals for threatening the people health by unsafe food stuff was more than 50 million US $. Of course, violation (*money*) or offence (*jail*) punishment are only a part of sanctions that due to the level of partaking in the incident, and the consequences, might be expected by the responsible ones. Meaning that almost by rule, follow very high damage chain request of the damaged and that what really means the largest punishment – losing the consumers' trust, reputation and the market position. The question of ethic and moral in connection to food safety, although very significant, this time we shall not comment as we have discussed on that topic fully, in the paper *"Catechism of seven principles HACCP concept"* [34]. So, all participants in the food production process and traffic must completely be aware of how important and responsible, and at the some time how honest and privileged job they undertake – a job that enables food for manhood, one of three life conditions. But food which is up to maximum correct and safe for human health as it is an imperative, condition of all conditions if we want to speak of any other aspect of food quality. A significant place and an extraordinary important role in this, as per our opinion, is the role and responsibilities of the government and their responsible departments and institutions. Of course besides the vision and global strategy, the government must create a suitable conditions, must work and do much more, to lead and to be the key initiator of numerous activities with an aim to create and develop national system for food safety improvement.

If carefully planned and structured activities result in good and firm foundation for better, more conscious and responsible work of all involved trades, all countries (including Serbia) could have a strong support and help from the authorities and very professional international institutions as FAO, WHO, CAC, etc.

These remarks are important mainly for the reason to once more (and who knows how many times up to now) we start paying closer attention to professional and scientific experts in this field. It has been underlined that the problem in Serbia is serious and a lot more work has to be done. The responsible government institutions need to be aware of poor work and results and take necessary steps to correct this problem.

We are the only country in the region not being able to establish a Food Agency, and to bring the Food Law We were not able, so far, to establish, accredit and authorize regional and national laboratories for food quality and safety research as well as further follow up. Also, we had no success in proclaiming an adequate law and sub-law regulations as a base for establishing an efficient and effective system for improving management for food stuff safety etc.

It is possible that the professional personal opinions are so strong, and interests and struggle in the area of inspecting responsibilities so challenging, that in this long period of time, we face almost paralyzed numbers of activities that have to be done. Without any pretence and already accepting the present situation in the country, as a part of our own professional responsibility, we would only like to point to some of very important issues which we have to deal with in the time ahead – as soon as possible:

- *establish national policy in relation with food safety;*
- *bring the Food Law; adequate standards and other following sub-laws;*
- *form authorized government institutions with clear and precisely defined responsibilities concerning management of food control and safety specially in human health;*
- *gather and engage all available sources in scientific and business capacities in the country;*

- *recognize the integrated managing approach related to food safety management;*
- *bring to a significantly higher level the existing inspection services by systematic and extensive education, trainings and knowledge upgrade;*
- *form specialized national laboratories for diagnosis and analysis (together with enabling essential infrastructure, adequate equipment and authorized executors);*
- *recognize the existing or, form new, regional centers for health monitoring, registering and updating information about incidents as well as concrete cases of diseases caused by unsafe food;*
- *establish efficient system for quick response to cases of serious and dangerous incidents caused by unsafe food;*
- *organize sufficient training and education of all involved in food safety issues (including consumers);*
- *establish efficient information and communication system available to entire population.*

3.5 Instead of a Conclusion

At the end of the last and the beginning of the twenty-first century, the human race is facing a numerous, very serious, global problems. Poverty and hunger are still present in many parts of the World (*mostly in Africa, parts of Asia, South and Central America*). The consequences of the weakness of the vulnerable and at the same time the lack of serious readiness of the most developed countries to help, are a huge concern. The problems with significant and rapid increase of the population, climate changes and alarming Arctic ice melting, as well as serious issues concerning food safety – need immediate attention and an effective action plan. Consequences of a rapid technological development, crazy race for profit and very often rude attitude of the most powerful industrial countries are very disturbing.

We are aware that the participants of the NATO Symposium (Cairo, Egypt, April 2010), or the readers of this article, are not in the position to make significant changes, and to a larger extent contribute to overcoming of the bigger part of listed global problems. But, all of us that have the privilege, that our own professional work and whole business activities in any way complete in the so called "food chain/nourishment" we have an obligation and moral task to do all so that the agricultural-nourishing goods would be correct, meaning in every sense safe for human health (*consumers*).

With this contribution, individually, modestly but joined of course more significant, that one of the global problems – *concerning the food safety and quality* – at least step by step and to a certain level overcomes, stop epidemics and concrete incident quicker and more successfully solve – having the less possible consequences. And something else: We always have to be aware that in life the winner is the one that reaches the goal last. So, if we all work to give meaning to it, and persistently and systematically improve food safety and quality – we shall build a firm foundation, and give a chance to all people. Life on our planet could last longer, many people can have better living conditions, feeling joy, happiness and enjoying way slower pace. We all shall try to provide, through our work and awareness of issues, an equal chance to each and every individual on this planet to have a healthy and long life, reaching the goal as late as possible.

References

1. WHO (2001) Surveillance programme for control of foodborne infections and intoxications in Europe. 7th Report – 1993–1998. WHO, Geneve
2. CDC (2007) Annual listing of FBD outbreaks by etiology. Centers for Disease Control and Prevention, Atlanta
3. WHO (2005) Annual report. WHO, Geneve
4. WHO (2006) Annual report. WHO, Geneve
5. WHO (2003) Annual report. WHO, Geneve
6. Radovanović R, Tomašević I, Tomić N (2006) Mesto i značaj HACCP koncepta u međunarodnoj trgovini prehrambenim proizvodima. Kvalitet 1–2:37–43
7. Radovanović R, Tomić N, Tomašević I (2005) Sistemi za upravljanje rizicima (HACCP/ISO 22000) obavezni deo poslovne strategije preduzeća prehrambene industrije. Zbornik radova XV Konferencija o kvalitetu (Jugoinspekt/JUQS/JUSK), 40–48. Čanj
8. Radovanović R, Tomašević I, Tomić N (2004) Troškovi i efekti unapređenja bezbednosti proizvoda industrije mesa – iskustva razvijenih zemalja. Zbornik radova XIV stručnog seminara o kvalitetu (Jugoinspekt/JUQS), 77–83. Petrovac
9. Radovanović R, Grujić R (2003) ISO 22000:200X: Novi korak u unapređenju bezbednosti hrane. Kvalitet 9–10:59–62
10. CDSC (2000) Annual report. CDSC, Washington, DC
11. ABS (2004) Australian Bureau of Statistics – Annual report. ABS, Kanbera/Sudney (2004)
12. Mead PS et al (2000) Food-related illness and death in the United States 5(5):607–625
13. Voogd E (2003) HACCP for food industry. In: Proceedings of the USDA/FDA Seminar, Belgrade, 19–23 May 2003
14. Botos M (2003) HACCP for food industry. In: Proceedings of the USDA/FDA Seminar, Belgrade, 19–23 May 2003
15. Grujić R, Radovanović R (2001) *Primjena HACCP u kontroliu namirnica,* u monografiji: Kvalitet i analiza namirnica. Ur. **Grujić, R.** *Univetzitet u B.Luci,* B.Luka 2001, RS, pp.511
16. Radovanović R (2003) Sertifikacija HACCP koncepta. Kvalitet 5–6:38–48
17. Radovanović R (2003) Sertifikacija HACCP koncepta – iskustvo razvijenih zemalja i naši nacionalni planovi. Zbornik radova XIII stručnog seminara o kvalitetu (Jugoinspekt/JUQS), 28–38, Petrovac
18. Rocourt J, Moy G, Vierk K, Schlundt J (2003) The present state of foodborne disease in OECD countries. WHO – Food Safety Department, Geneve, 2003
19. FAO/WHO (2006) Food safety risk analysis – a guide for national food safety authorities. Food and Nutrition, Paper 87. FAO, Rome
20. UN Department of Economic and Social Affairs (2006) The Millenium Development Goals Report. Goal 1: Eradicate extreme poverty and hunger. UN Department of Economic and Social Affairs, New York
21. FAO (2007) Annual report. The spectrum of malnutrition. FAO, Rome
22. World Food Programme (2007) Food aid to save and improve lives. FAO, Rome
23. FAO (2007) Water and food security. FAO Aquastatistics, Rome

24. FAO (2006) Statistical databases. FAO, Rome
25. CAC (1969) Recommended international code of practice – general principles of food hygiene, including annex on hazard analysis and critical control point (HACCP) system and guidelines for its application – CAC/RCP-1 (1969), Rev. 4, 2003
26. WTO (1998) Understanding the sanitary and phytosanitary measures agreement. WTO, Geneva
27. FAO/WHO (1994) The Global Forum of Food Safety Regulators. Proceedings. Marrakesh, 28–30 January 1994
28. Radovanović R, Tomašević I, Tomić N (2006) Uloga i značaj HACCP koncepta u međunarodnoj trgovini mesom i proizvodima od mesa. Tehnologija mesa 1–2:8–19
29. FAO/WHO (1991) Report of the FAO/WHO conference on food standards, chemicals in food and food trade. Licom 91/22. WHO, Rome, 18–27 Mar 1991
30. Garret SE, Jahncke LM, Cole EA (1998) Effects of Codex and GATT. Food Control 9(2–3):177–182
31. Caswell AJ, Hooker HN (1996) HACCP as an international trade standard. Am J Agric Econ 78:775–779
32. OECD-FAO (2006) Agricultural outlook 2006-2015. FAO, Rome
33. WHO (2007) International travel and health – 2007 Edition. WHO, Geneve
34. Radovanović R, Tomašević I (2006) Katehizam 7 principa HACCP koncepta: 7 smrtnih grehova i 7 nebeskih vrlina u odnosu na bezbednost hrane. Kvalitet 9–10:37–4

Chapter 4
Solidarity Principle as a System Tool for Mitigation of Flood Impacts

Scope and Effectiveness in Western Balkan Countries: An Issue for Investigation

Slavko Bogdanovic

Abstract Floods, in Western Balkans, like in other parts of EU and the world, caused by natural (*e.g.* recently caused more often by climate changes) and anthropogenic reasons, could be seen nowadays frequently and of the size not so often seen in the past. This uncontrollable natural phenomenon impacts human life and health, economy, and environment and threatens food security in affected areas. This paper contains a concise review of solidarity principle application in the EU policies, some indications on current global trends, and state of things in the Western Balkan countries, relevant for food security. The paper is not the result of a focussed research into the subject matter, but rather a report based on author's side observations collected during researches into other water (policy and law) issues. Thus, it should be taken as a kind of preliminary research or an outline (or synopsis) for further comprehensive and in-depth research into the existing systems of response on emergency situations caused by waters in the Western Balkan countries and interconnected flood, and food security policy and legal issues.

Keywords Western Balkans • Floods • Environment and security • Solidarity principle • Food security

4.1 Introduction

Floods in Europe, in all their forms of appearances (fluvial, flash, urban, and flooding of coastal areas) are a natural phenomenon having the potential to undermine efforts of the European Union towards sustainable development and

---

S. Bogdanovic (✉)
The Business Academy, Independent and Non-State University of Novi Sad, Novi Sad, Serbia
e-mail: slavkob@nadlanu.com

M. Hefnawy (ed.), *Advances in Food Protection*, NATO Science for Peace and Security Series A: Chemistry and Biology 1, DOI 10.1007/978-94-007-1100-6_4,

competitive economy [1], p. 7. They cannot be prevented, because of the uncontrollable natural occurrences causing them (rainfall and sea levels rising). The risk from floods is potentially enormous, having in view the number of people living and assets located in the areas at risk from flooding. Floods pose a potential threat to human health and life, economic assets, including private housing, public infrastructure, commercial and industrial enterprises, and the environment [2], p. 2. Besides direct damage from floods, indirect damage could also be extremely high, in terms of clean-up costs, loss of client and markets, loss due to disruption of production (which may lead to closure of businesses or SMEs) [1], passim. The damage in agriculture can comprise damage to crops and cattle but also damage to the machines (equipment, stocks) and buildings [2], p. 48. Extensive damage, threat of repeated floods and withdrawal of insurance coverage could make properties impossible to sell [1], p. 9.

In future, the risk of floods is expected to increase in terms of scale and frequency, due to climate change (higher intensity of rainfall and rising of sea levels), inappropriate river management, and construction in flood prone areas. An increase of vulnerability is also expected due to the population and assets located in flood risk zones [2], p. 2.

In the period between 2002 and 2008, Europe had over 100 major floods, some of which were categorized as catastrophic (in the Danube and Elbe River Basins in 2002). European Commission had undertaken comprehensive activities which eventually led to adoption of the Directive 2007/60/EC on the assessment and management of flood risks. That Directive, fill in the gap in EU legislation which was left open after adoption of the Water Framework Directive 2000/60/EC [4], which did not set regulation of flood management risks as its objective. The Directive 2007/60/EC addresses the issue of the impact of climate change on flood risks [Article 4.2] and provides for establishment of a framework for the assessment and management of the flood risks, aimed at reduction of the adverse effects for human health, the environmental, cultural heritage, and economic activities [Article 1]. Jointly taken, by regulating water quality management and flood risk management respectively, two directives make a coherent concept of integrated river basin management [2], p. 7. Management the risk of floods within this concept includes several elements, *i.e.* prevention, protection, preparedness, emergency response and recovery, and lessons learned [5], pp. 3–4. This concept has been based *inter alia* on the respect of the principle of high level of environmental protection and improvement of the quality of environment in accordance with the principle of sustainable development [6]. As a consequence of new approach to copping with floods in EU, flood maps were developed to provide information about hazards, vulnerabilities, and risks and are an irreplaceable tool for implementation of the necessary preventive and preparedness measures [7].

This clear European picture can only be supported by findings on the global scale of the UN General Assembly, stating that progress in attaining hunger and poverty reduction targets (Millennium Development Goal 1) "has been dealt a severe setback, and a reversal progress has taken place [8], p. 5." In particular, it has been pointed out that small farmers are highly vulnerable to shocks, including *inter alia* price volatility, natural hazards, and climate variability [Op. cit., p. 8]. Efforts for developing of a strategy aimed at sustainable agriculture and improving food security are accompanied with efforts for adaptation of water resources management to climate change, which has been seen, together with land, as critical to "sustainable development, especially in context of worsening food security and malnutrition" [9], p. 6. In this contexts, an integrated approach to flood management would comprise development of land and water resources in a river basin and would require interaction between different professions, state authorities, and sectors of society [10], p. vii.

4.2 Solidarity Principle in EU

The EU response to floods has been based on the solidarity principle, which was expressed in a Commission Communication from 2002 in the following way:

> "We are a community of peoples on the path to closer union. It is only right and natural therefore that citizens, member States and Community Institutions feel a spontaneous urge to show their sympathy for the victims of the floods through practical gestures of financial solidarity in particular [11], p. 3."

In the same year, the EU Solidarity Fund (EUSF) was established [12], with the aim to enable Community to respond rapidly, efficiently, and in a flexible manner to emergency situations [Article 1], which are inter *alia major* natural disasters (*i.e.* such disasters resulting in at least one of the Member States in damage estimated either over three billion in 2002 prices, or more than 0.6% of its GNI). EUSF may be mainly mobilized in case of serious repercussions on living conditions, natural environment, and economy [Articles 2.1, 2.2]. The aim of EUSF is to complement the efforts of the affected State (or the region) to carry out emergency operations, which could comprise:

- Immediate restoration of works of infrastructure and plants in the fields of [*inter alia*] water and waste water;
- Provide temporary accommodation and funding rescue services to satisfy the immediate needs of population;
- Immediate securing of preventive infrastructures and measures for immediate protection of cultural heritage;
- Immediate cleaning-up of disaster-stricken areas, including natural zones [Article 3.2].

EUSF may only intervene for emergency operations, and not for phases preceding an emergency [12].

The principle of solidarity has been considered very important in the flood risk management. The Member States are encouraged to seek for fair sharing of responsibilities when deciding on joint measures concerning flood risk management of shared watercourse. Flood risk management plans should be developed in such a way that flood protection measures do not compromise the ability of other upstream or downstream regions or Member States to achieve the same level of protection as the region or State developing the plans. The strategy considered to be appropriate consists of a three-step approach – retaining, storing and draining [waters].

EUSF does not compensate for to individuals, for private loses or for damage covered by insurance [5], p. 5. However, in the EU Member States, compensation for material damage to individuals caused by floods can be up to 100%. Moreover, compensation could include previously awarded aid the effects of which were "washed away." Even indirect damage caused by floods (in terms of production delays due to electricity cuts, difficulties in delivering products, due to blockage of certain transport route) can be fully compensated to individuals if a clear causal link between damage and flood can be established [11], p. 8.

The possibility for compensation of individuals for damage suffered from floods is not based on the solidarity principle but on the rules applicable to state aid. The state aid control is an old EU policy, deemed by the Commission to be a crucial element of free and fair competition on the single EU market. It has two regulatory dimensions, determining individual grants of aid and aid schemes which establish national frameworks of subsidisation [13]. The state aid rules of the EU are rules regulating support of the EU to rural development in the Member States through the European Agricultural Fund for Rural Development (EAFRD), but under conditions preventing the Member State's support to result in market distortions. The state aid provided to individuals, who suffered damage from floods has been exempted from those EU rules [14], p. viii.

4.3 Solidarity Principle in the Western Balkan States

The principle of solidarity in emergency situations was in earlier times the basic one in the national legal frameworks of Western Balkan countries (in particular in the Socialistic Federal Republic of Yugoslavia – SFRY) which provided high level of protection and security of population and material assets in cases on natural (and industrial) disasters/accidents. The response of society to the natural and other major disasters was regulated in great detail by the laws of the republics and autonomous provinces – constitutional elements of federation. The notion "natural and other major disasters" comprised *inter alia* atmospheric storms, floods, torrents, accumulation of ice in the watercourse and breaking of dams on watercourses [15].

Protection against natural disasters and other major disasters was based on the constitutional principle of solidarity which enabled republics and autonomous provinces to adopt their laws and regulate all financial aspects of protection from such occurrences and subsequent rehabilitation measures. They adopted their laws on use of funds for rehabilitation and protection against natural disasters. The funds were designed as budgetary ones, and grants to affected individual owners (of land, buildings, and other property) were available through municipalities, after their having obtained assessment of damage done, in a period of 60 days after disaster. Individuals were eligible to be fully compensated for material damage as well as municipalities for *e.g.* infrastructure. Such role (obligation) of federal units was determined in their constitutions. In that system the insurance plays not so important role.

The new Western Balkan states originated in the territory of former SFR Yugoslavia during 1990's and 2000's are on their own ways towards association with EU. Their constitutions contain sometimes provisions that would be the most general legal ground for building-up relevant primary and secondary legislation that would provide for response of state to natural disasters (floods, *inter alia*) in accord with EU standards. However, it seems (this is a hypothesis not based on systematic comparative investigation of the constitutional and legal system of Western Balkan countries) that solidarity principle has not clearly been established. To be fair, it should be mentioned that the solidarity principle has been introduced in new Law on Emergencies of the Republic of Serbia [16], in spite of the fact that there is no constitutional ground for imposing it. In that Law, the solidarity principle was constructed in a sentence declaring that "everyone shall take part in protection and rescue [operations] in accord with its own capabilities and skills [Article 5(b)]." The Law *inter alia* provides for rescue and aid assistance in operations undertaken for protection and rescue of human lives, material assets and environment, and mitigation and rehabilitation of immediate consequences of extreme natural occurrences and other disasters. The assistance may be aimed at establishment of necessary conditions for citizen's lives in the affected area [Articles 3(4), 3(5)].

Financial mechanisms provided by this Law do not contain clear obligation of the state, province, or municipality to compensate individuals for damage suffered from floods. Such compensation (that would be a kind of state aid) may appear as *ad hoc* interventions of the State in the form of provision of material assistance for mitigation of consequences of *e.g.* floods, preventing new damage to occur and securing normal conditions for life in the affected area, as well as of satisfying other needs for protection and rescue in accord with other legislation [Article 132]. It is not likely that assets of individuals (in particular in agriculture) are protected by the system in the way that compensation for material damage caused *e.g.* by floods (or hail) would be provided.

Consequentially, if this hypothesis is right, in situation of repeated and extreme flood occurrences (that should be expected due to the climate change) individual farmers, owners, and tenants of agricultural land would be in a vulnerable situation and not able to play their role in food security chain. At the scale of the Western Balkans region, a long-term lack of clear system of compensation for material damage in agriculture caused by floods would worsen living conditions of large part of population, and probable cause failing of the countries to achieve MDGs. This is more likely to be expected if the insurance market is not developed so far and able to offer insurance schemes that would be acceptable to owners and tenants of agricultural land.

4.4 Closing Notes

In conclusion, it could be noted that there is a global concern in regards of climate change and connected changes in water cycle and impact of those changes on agriculture and food security. Accordingly, policy and legal measures are developed and recommended to be adopted to cope with those unpredictable and uncontrollable natural phenomena. Expecting the problem to be even more serious in future (due both the climate changes but also to anthropogenic causes), the European Union has developed its own system of responses to the flood risks through development and implementation of the flood risk assessment and management approach in the framework of the concept of integrated water management. As a core part of the system, the solidarity principle was developed and institutional measures were deployed with the aim of reacting rapidly, efficiently, and in a flexible manner. The system of state aid has been structured in such a way to provide possibility for individuals, particularly in agriculture, who suffered damage (direct and indirect) by floods, to be compensated for material damage. In that way, producers in agriculture are protected from adverse impacts of floods, and the level of their vulnerability has been decreased. Such approach undoubtedly should be seen as a concept supporting the food security in EU.

All the Western Balkan countries are in the association process the goal of which is their membership in EU. In regards of flood risks, all of them (or at least those countries that originate in the territory of former SF Yugoslavia) have a legacy of a comprehensive solidarity principle which was established by their socialistic constitutions. On the basis of socialistic solidarity principle, full compensation for material damage caused by natural disaster (*e.g.* floods) had been provided to individuals, but also to municipalities (for infrastructure). In changed social environment, the Western Balkan countries are faced with emergency situations (repeated floods) caused by changes in nature (climate). Their systems of responding on emergencies are under development, and most probable having not adequate concepts of compensation for damage suffered by small farmers (land owners and tenants). These issues have not been investigated, and the conclusions here are hypothetical ones. A comparative research into the constitutional grounds and respective current national legislation would show whether legal and institutional frameworks in Western Balkan countries are supportive to food security of nations and the region. Such investigation would identify the gaps in national legal systems and provide a firm basis for conclusions on and recommendations of measures to be undertaken in line with broader international policy requirements in this regards.

The new, supranational (*i.e.* EU) paradigm concerning flood risks assessment and management that has been evolving, combined with state aid schemes and state aid support to individuals, aimed at rural development should be the pathway for development of national legal systems of Western Balkan countries, harmonized as much as possible with the Community *acquis* during the transition period (i.e. period of approximation of national legislation to the EU legislation).

References

1. Commission of the European Communities (2006) Commission staff working document – Annex to the proposal for a directive of the European parliament and of the council on the assessment and management of floods, impact assessment {COM(2006) 15 final}. Brussels, 18 Jan 2006 SEC(2006)66
2. Commission of the European Communities (2006) Proposal for a directive of the European parliament and of the council on the assessment and management of floods – Explanatory memorandum, {SEC(2006)66}. Brussels, 18 Jan 2006 COM(2006) 15 final, 2006/0005(COD)
3. Lamothe DN, Neveu G, Görlach B, Interwies E (2005) Evaluation of the impact of floods and associated protection policies – final report. European Commission, DG Environment
4. Directive 2000/60/EC of the European parliament and of the council of 23 Oct 2000 establishing a framework for community action in the field of water policy, OJ L 327/27, 22 Dec 2007
5. Commission of the European Communities (2004) Communication from the commission to the council, the European Parliament, the European Economic and Social Committee and the Committee of the Regions – Flood risk management, flood prevention, protection and mitigation. Brussels, 12 July 2004, COM(2004)472 final
6. Charter of Fundamental Rights of the European Union (2000/C 364/01) OJ C 364/1, 18 Dec 2000
7. EXIMAP – European exchange circle on flood mapping. http://ec.europa.eu/environment/water/flood_risk/flood_atlas/pdf/handbook_goodpractice.pdf. Accessed 19 July 2010
8. UNGA (2009) Agriculture development and food security – report of the Secretary-General, for the sixty-fourth session, A/64/221, 3 Aug 2009
9. UNGA (2010) Water, climate change and disasters – note by the Secretariat, for the sixty for session. Agenda item 53(a)-sustainable development: implementation of Agenda 21, the programme for the further implementation of Agenda 21 and the outcomes of the World Summit on Sustainable Development, A/64/695, 4 Mar 2010
10. WMO: ASPECTS JURIDIQUES ET ISTITUTIONNELS DE LA GESTION INTÉGRÉE DES CRUES; PROGRAMME ASSOCIÉ DE GESTION DES CRUES; Organisation météorologique mondiale, Programme associé de gestion des crues, IWLRI, PARTENARIAT MONDIAL POUR L'EAU; OMM-No 997, Geneva 2006 [Edition in Serbian: Yugoslav Association for Water Law, Novi Sad 2006]
11. Commission of the European Communities (2002) Communication from the commission to the European parliament and the council – The European community response to the flooding in Austria, Germany and several applicant countries – a solidarity based initiative. Brussels, 28 Aug 2002, COM(2002) 481 final
12. Council Regulation (EC) No. 2012/2002 of 11 November 2002 establishing the European Union Solidarity Fund, OJ L 311/3, 14 Nov 2002
13. Michelle C(2000) From soft law to hard law?: discretion and rule-making in the commission's state aid regime. European University Institute Robert Schuman Centre for Advanced Studies, RSC No. 2000/35
14. Council Regulation (EC) No. 1698/2005 of 20 Sept 2005 on support for rural development by the European Agricultural Fund for Rural Development (EAFRD), OJ L 277/1, 21 Oct 2005
15. Bogdanovic S (2009) Protection against detrimental effects of waters: the legal system of Serbia. In: WMO: legal and institutional aspects of integrated flood management – case studies. WMO – World Meteorological Organization, APFM – Associated programme for Flood Management, GWP – Global Water Partnership, pp 51–85
16. Official Gazette of the Republic of Serbia (2009) No. 111/2/2009

Chapter 5
Higher Risk of Health Problems for Radio-Contaminated Food Consumption by Human Population Suffering Protein-Calorie Malnutrition and Environmental Chemical Pollution

Hamed M. Roushdy

Abstract The incidence of contaminants in many food commodities including:

- Pathogenic organisms, parasites, insects, chemical residues, and radio-nuclides contribute seriously to the high rate of risk imposed on a vast sector of population suffering protein – calorie malnutrition.
- Radionuclide contaminants in different food stuffs originate from various sources including nuclear test explosions, nuclear accident, nuclear wastes, and radioactive material encountered in technologies of nuclear fuel cycle.
- Both internal radioactive contamination as well as protein calorie-malnutrition exert damaging effects on the human genome and retard the repair mechanism in human living cells. The synergistic effect of both incidences contributes to the aggravation of cellular damage and suppression of repair mechanisms.
- It is ethically recommended to revise the safety level for internal radio-contamination in human subjects suffering health problems including protein calorie malnutrition, parasitic infestations, and heavy environmental pollution prevailing in many developing nations.

Keywords Radioactive contamination • Protein-calorie malnutrition

5.1 Environmental Pollution

Pollution of man and his biosphere is formulating a steadily rising threat to maintenance of life in our globe. As presented in Fig. 5.1, different sources of environmental pollution are shown which play an ultimate role in damaging DNA Human Genetic Material.

H.M. Roushdy (✉)
Department of Radiation Biology, National Center for Radiation Research and Technology, Atomic Energy Authority, Nasr City, Cairo, Egypt
e-mail: roushdy@excite.com

M. Hefnawy (ed.), *Advances in Food Protection*, NATO Science for Peace and Security Series A: Chemistry and Biology 1, DOI 10.1007/978-94-007-1100-6_5,

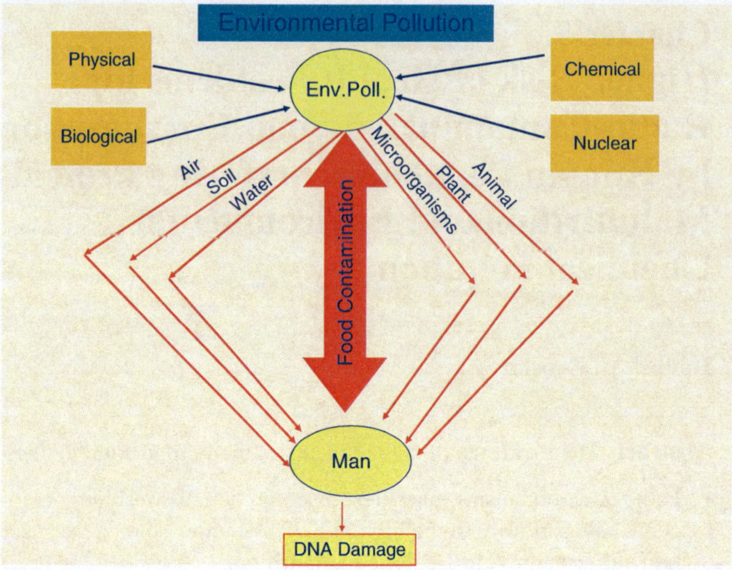

Fig. 5.1 Environmental Pollution

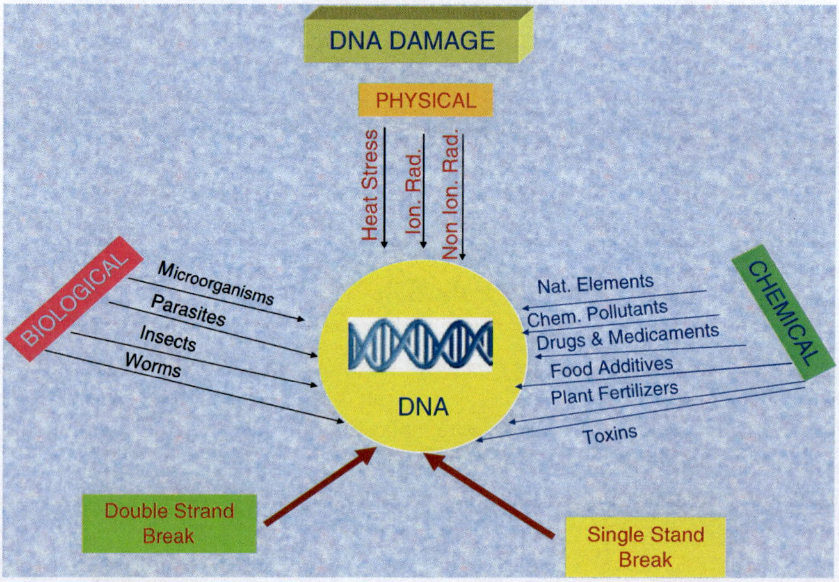

Fig. 5.2 DNA Damage

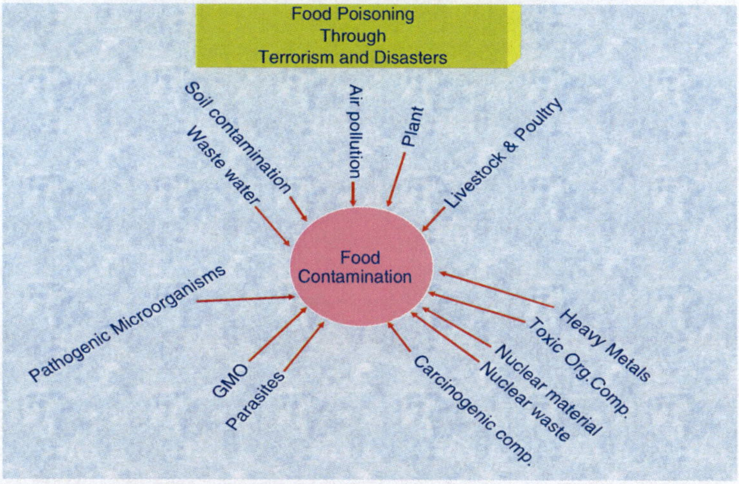

Fig. 5.3 Food Poisoning Through Terrorism and Disasters

Human food, animal feed, and plant fertilizers are presented to be good carriers for environmental pollutants and food contaminants: Figs. 5.2 and 5.3.

Many developing nations are subjected to environmental conditions encountering higher heat stress, shortage in water and adequate food supplies, low plant and animal production, high rate of food losses, high incidence of bacterial, viral and parasitic cross-infection, besides many other socio-economic problems imposed by poverty and illiteracy resulting in exhaustive expenditures on importation of food.

5.2 Environmental Radio-Activity

It is known that all environments have a natural radio-activity of their own. Natural sources of radiation , over which little or no control can be exerted, are cosmic rays, radio-isotopes naturally generated in air envelope, radiations from earth and building material as well as radioactive substances regularly formed as natural constituents of the living body.

On the other hand, artificial radioactivity can be attributed to radioactive material produced by nuclear explosions, radioactive wastes, and radioisotopes of all kinds discharged into the environment.

A list of the major sources of environmental radio-contamination is presented by Fig. 5.4.

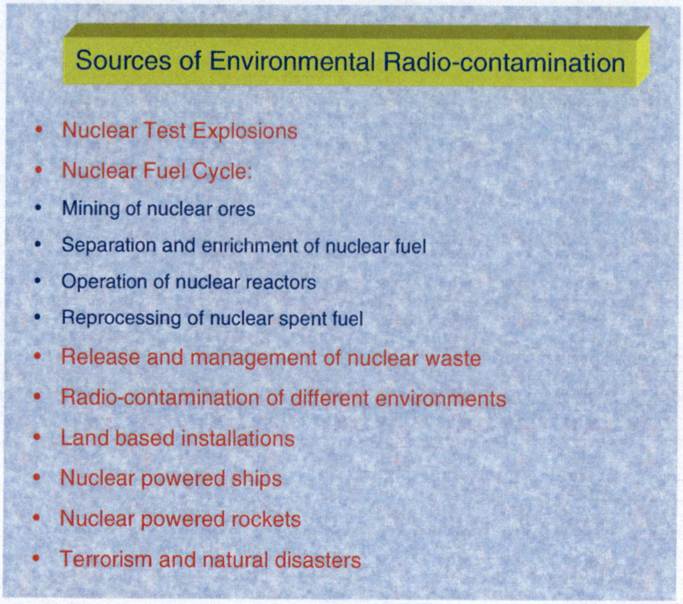

Fig. 5.4 Sources of Environmental Radio-contamination

5.3 Internal Radioactive Contamination in Human Subjects

Subsequent to the entry of a radioactive substance into the human body, a part of the radioactive contaminant is metabolized in the organ of intake whereas the rest gets eliminated by the body excretory processes. The metabolized part remains internally deposited for a relatively longer duration and leaves the site of intake by dissolution to the blood. Upon absorption, the radioactive contaminant follows its specific biokinetics in the human body leading to deposition in other organs and excretion over prolonged periods of time. The biokinetic behavior of a particular radionuclide in the human body determines its source and target organs/tissues. In case of actinides, Th., U, Pu, Am etc. these, upon uptake, largely distribute in the skeleton and liver.

Mathematical models are used to describe the various processes involved in internal deposition and retention of radionuclide and the associated radiation doses received by various organs and tissues in the body. Such models are developed from knowledge and assumptions about the behavior of various radionuclides in the body or in specific organs. In such models, the body is viewed as a series of compartments into which the radioactive materials enter and exit at various rates, ultimately being removed from the system by some form of excretion, by radioactive decay or both.

5.4 Models of Intake of Radionuclides in the Human Body

Information on the behavior of radionuclides ingested in the human body is important for assessing the mode of distribution, pattern of deposition, rate of release, and for optimizing treatment schedules of decontamination. The choice of radio-measurement technique is determined by several factors: type of radiation emitted by the radionuclide, the metabolic behavior of the contaminant, its retention in the body organs, its biological clearance, and its radioactive decay. For instance, radionuclides that emit X-rays or Gamma-rays can be measured by whole body counter while alpha- and beta- emitters must be measured by bioassay techniques on excreta samples: urine and feces.

When a radionuclide is ingested through food intake, the amount of radionuclide excreted soon after ingestion represents the amount of radionuclide unabsorbed all over passage through the gastrointestinal tract. The fraction of ingested radionuclide that is absorbed enters the blood stream and is progressively deposited in various organs. A portion of the radionuclide entering the blood stream is subsequently excreted from the liver into the gastrointestinal tract via the bile and is then excreted from the body through feces (Fig. 5.5).

Any radionuclides remaining within the principal organ of intake, the gastrointestinal tract, will irradiate that organ and its tissues. The rate at which the radionuclide leaves the site of intake through dissolution and deposition into the blood, depends on

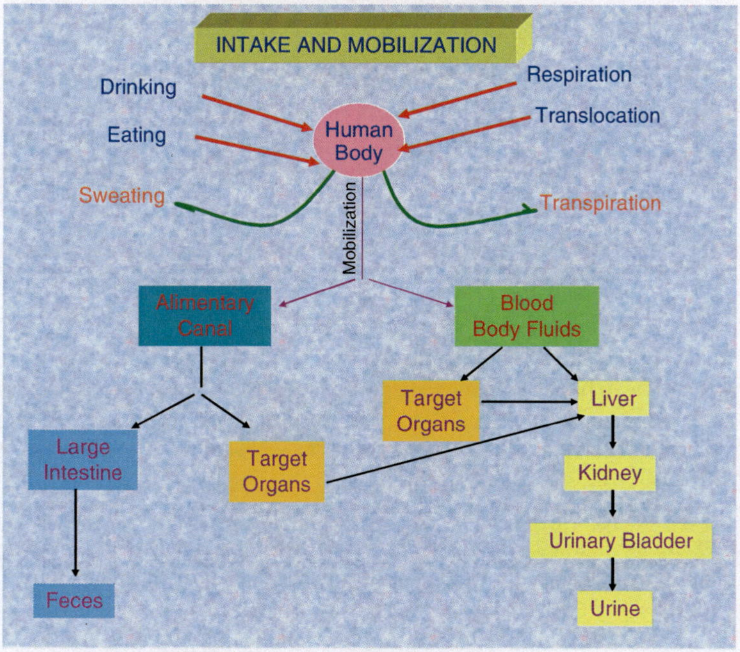

Fig. 5.5 Radioisotopes intake and mobilization

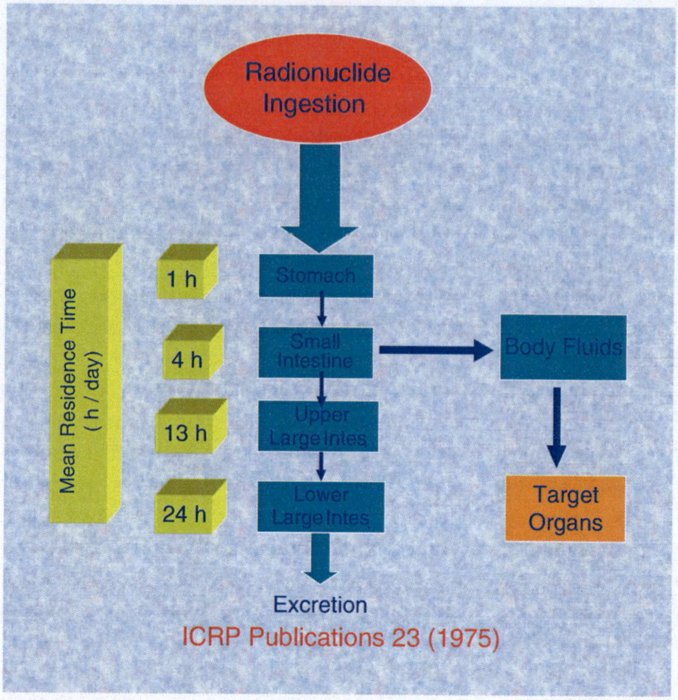

Fig. 5.6 Radionuclide Ingestion

the physical and chemical properties of the material itself. Radioactivity measurement depends mainly on the emission and the solubility characteristics of the radionuclides concerned. Among these: the type of radiation emitted by the radionuclide, the metabolic behavior of the contaminant, its retention in the body organs, its biological clearance, and its physical radioactive decay, as well as the availability and convenience of the appropriate measurement facilities (Figs. 5.6–5.8).

5.5 Biological Effects of Radiation

It is known that all ionizing radiations corpuscular (alpha, beta, protons and neutrons) or electromagnetic (X and Gamma rays) at sufficient levels are toxic to living organisms, and in higher dosage are lethal to biological systems and living organisms. However, the radioactivity of different biological targets shows wide variations.

In living cells, radiation exposure at sufficient dose levels, results in a complex of initial changes which may exert both short and long-term consequences. When cellular damage occurs and is not adequately repaired, it may prevent the cell from reproducing or even surviving, or it may result in a still viable, but modified cell. These outcomes have different implications on the living cell, tissue, organ, or the body as a whole.

PATHWAYS OF RADIONULIDES ACROSS HUMAN FOOD AND ANIMAL FEED

Carrier	Critical Radio-nuclides	Pathways	Critical Target Organs
Milk	Sr-89, Sr-90, I-131, Cs-134 and Cs-137	Grass-Cow milk and Milk products	Thyroid gland-GI tract and whole body
Meat	Cs-134 and Cs-137	Grass - Meat	Whole body, GI Tract and Muscles
Other Food	Sr-89, Sr-90, Cs-134 and Cs-137	Fish, Vegetables and Fruits	GI tract and Whole body
Vegetation	Sr-89, Sr-90, Zr-95, Nb-95, Ru-103, Ru-106, I-131, Cs-134, Cs-137, Ce-141 and Ce-144	Grass-Cow milk, Leafy vegetables and Fruits	Thyroid gland-GI tract and whole body

Fig. 5.7 Pathways of Radionulides Across Human Food and Animal Feed

HALF-LIVES

Radioactive Elements	$T_{1/2}$	
Tc-99m	Hours	6.0
Na-24		15.0
Au-198	Days	2.6
Xe-133		5.27
I-131		8.05
P-32		14.3
Cr-51		27,8
Fe-59		44.6
Ca-45		162.6
Zn-65		245.0
Cs-134	Years	2.07
Co-60		5.26
Sr-90		29.0
CS-137		30.0
Am-241		462
Rd-226		1620
C-14		5730
K-40		1280×10^6
U-238		4510×10^6

Fig. 5.8 Radioisotopes Half-lives

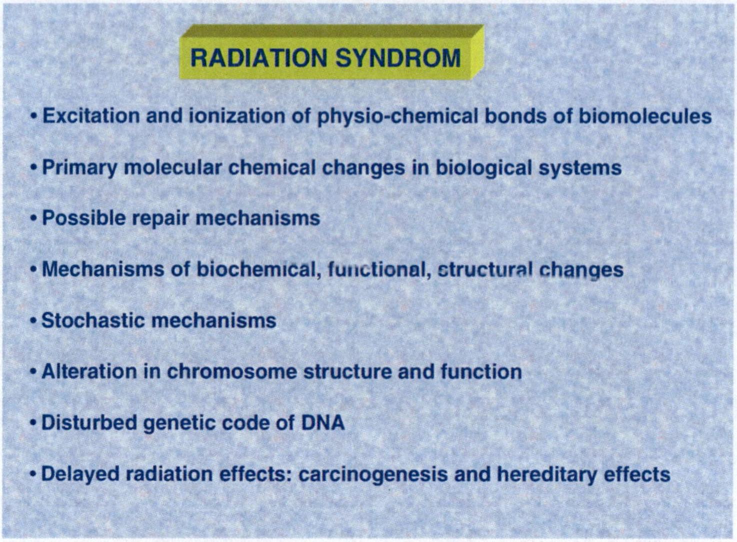

Fig. 5.9 Radiation Syndrom

If the number of cells damaged or lost in a tissue is large enough, there will be a detectable harm, as reflected in the impairment of tissue function and structure. The severity of the harm increases with the radiation dose level. This type of effect is called "deterministic."

If the loss of cells can be compensated by biological repair and repopulated, the effect will be of a relatively short duration. When the irradiated cells are only modified rather than killed, certain cells may then be able to produce clones of modified daughter cells which may cause after a prolonged and variable delay of time, malignant cancer cells. This kind of effect is called "stochastic." When radiation damage occurs in a cell whose function is to transmit genetic information to later generations, this type of stochastic effect is referred to as "hereditary." (Fig. 5.8).

Deoxyribonucleic acid (DNA), the genetic material of the living cell, is the most important target for radiation interaction in living organisms. The DNA polymer is the source of information that passes from a cell to its descendents. The DNA is known to be damaged by radiation through a direct effect of ionization in the DNA structure and an indirect effect due to active chemical radicals produced by radiation in the vicinity of the DNA. The diffusion of the radicals to the DNA results in chemical changes (Fig. 5.9).

5.5.1 Deterministic Effects

High radiation doses can deplete the living tissues sufficiently to cause functional failure. The severity of the deterministic effects depends on the radiation dose level and the varying susceptibility of the target organ. Examples of the deterministic effects include the induction of temporary or permanent sterility in gonads, depression of the effectiveness of the blood forming system leading to a progressive decrease in the number of blood cells, epilation, loss of skin surface, induction of opacities in the crystalline lens of the eye (cataract), as well as visual impairment and inflammation in certain body organs.

5.5.2 *Stochastic Effects*

Radiation induced changes in the cell genetic code (mutation) can be lethal to cell progeny or can result in viable, but modified cells. When these cells belong to gonadal cell lines (ova or sperms), the change is expressed as hereditary.

Cancer initiation involves loss of control of the cell reproduction cycle and its differentiation processes.

5.5.3 *Hereditary Effects*

If the change of the genetic code occurs in the germ cells (sperms or ova) or in the germ mother cells, this effect is transmitted and may result in hereditary disorders in the descendents of the exposed individuals. Such disorders may cause gross loss of organ function, anatomical changes, and early death. Some radiation – induced dominant mutations can be passed silently through several generations and then suddenly cause their damaging effect (Figs. 5.10 and 5.11).

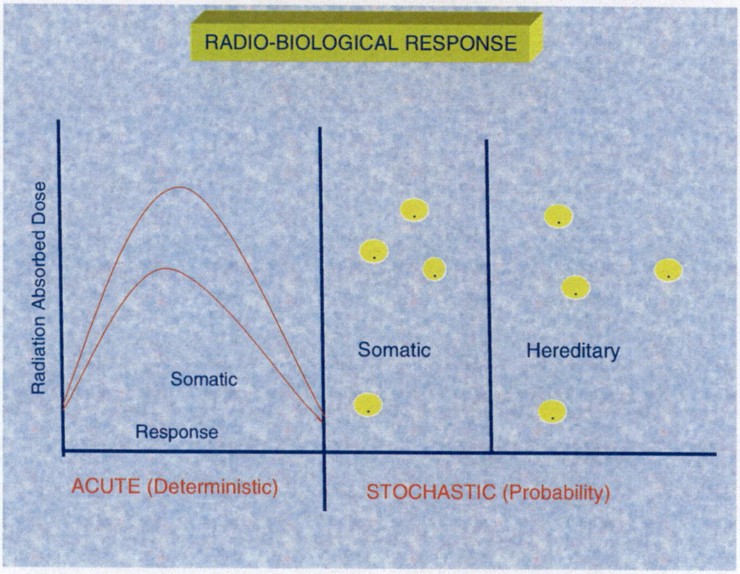

Fig. 5.10 Radio-Biological Response

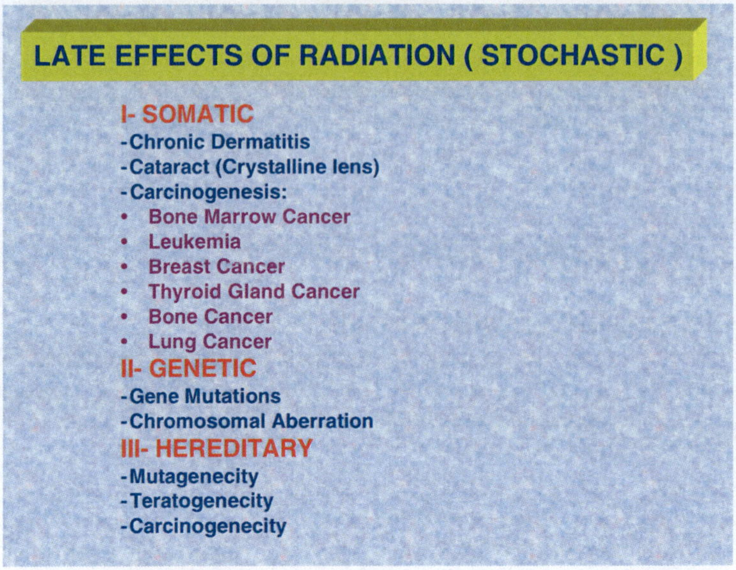

Fig. 5.11 Late Effects of Radiation (Stochastic)

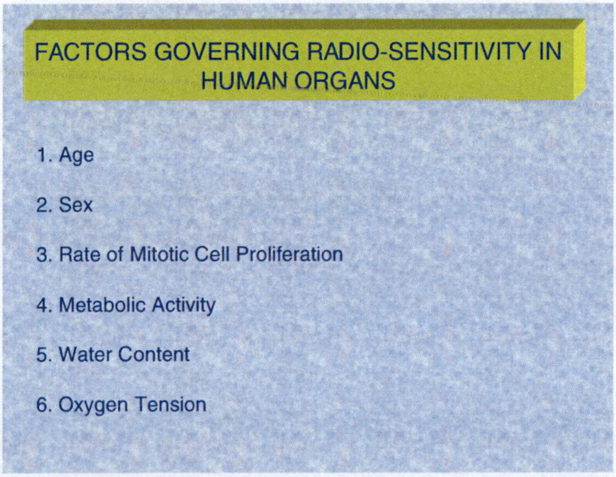

Fig. 5.12 Factors Governing Radio-Sensitivity In Human Organs

5.6 Relative Radio-Sensitivity of Biological Targets

There are many factors influencing the radio-sensitivity of living organisms and tissues. Among these are: (Fig. 5.12)

- On the Cell and Tissue Level: The rate of cell mitotic activity, the stage of mitosis, the degree of cell differentiation, the rate of metabolic activity, the water content, the rate of cell repair, tissue regeneration, and the rate of oxygenation.
- On the Organism Level: The age, sex, sporulation (in case of microorganisms), and individual variation.
- On the Species Level: In general, mammals including man show the highest rate of radio-sensitivity amongst the animal kingdom, whereas unicellular organisms show the highest radio-resistance.

5.7 Recovery Processes

Attempts of repair and regeneration are an early response of living cells to any form of injury. The rate of recovery processes in irradiated biological targets is affected by many physical and biological factors. Among the biological factors are the rate of cell proliferation, rate of blood supply, microbial infection, nutritional status, water content, species, age, sex, type of organ or tissue, and individual variations.

5.8 Environmental and Health Problems Boosting Potential Risk of Radiation Injury in Developing Nations

Consumption of radio-contaminated food commodities by man or animal underlie potential health problems both on the somatic and genetic levels.

In certain developing nations, mainly in the arid zones, a complex of environmental and health problems plays a synergistic role for potentiation of the deleterious effect of radio-contaminated food intake. Among these are the higher risk of genetic damage as affected by heat stress, environmental chemical pollutants, immune deficiency against cross infection, reduced rate of biological repair of genetic damage as affected by protein-calorie malnutrition and hereditary disorders (Fig. 5.13).

5.9 Ethical Bio-Safety Measures Undertaken to Control Exposure to Environmental Radioactivity in Developing Nations

It is known that an important goal of bio-ethics is risk assessment encountering careful analysis and prediction of risk.

Extensive periodical monitoring surveys on the radioactivity levels of the local environment encountering extensive sampling of ecological masses and food chains components: air, water, soil, plant, and animal should be undertaken.

Adjustment of acceptable radioactivity dose limits for food should be drafted taking into consideration the local food habits and average food intake governed by the socio-economic status of the population. Particular attention should be given to

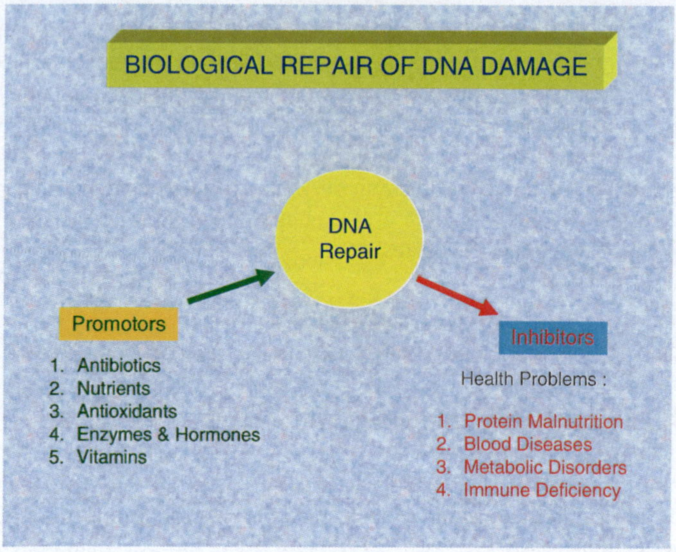

Fig. 5.13 Biological Repair of DNA Damage

the body burden of low income communities suffering stress of parasitic infestations and protein calorie malnutrition syndromes. In all cases, the ICRP ALARA concept as well as the limits adopted by the concerned UN Organizations should be referred to.

5.10 Accidental Intervention Levels Adopted in Egypt for Food Radio-Contamination Post-Chernobyl Nuclear Power Plant Accident (April 1986)

General Considerations:

- Elaborate monitoring of radioactivity levels of different ecological masses, confirmed freedom of Egyptian environment from direct post-accidental Chernobyl residual radioactivity.
- Since Egypt, as the case in many developing nations, is still importing large amounts of food commodities for both human and animal consumption, acceptable intervention levels for radioactivity in imported food shipments were established and rigidly enforced.

Adjustment of acceptable radioactivity dose levels were drafted in view of the following basic information:

- Local food habits and average annual food intake under the prevailing socio-economic status.
- Radio-sensitivity and radio-body burden of low income communities suffering other stresses of parasitic infections and protein calorie malnutrition syndromes.
- ICRP ALARA concept for minimizing radiation exposure to levels "as low as reasonably achievable."
- Foreign trade policies and acceptable radioactivity dose limits adopted by other countries for international food trade (Figs. 5.14–5.16).

Fig. 5.14 Egypt's Accidental Intervention Adopted for Food Radio-contamination Post Chernobyl Nuclear Power Plant Accident April 1986

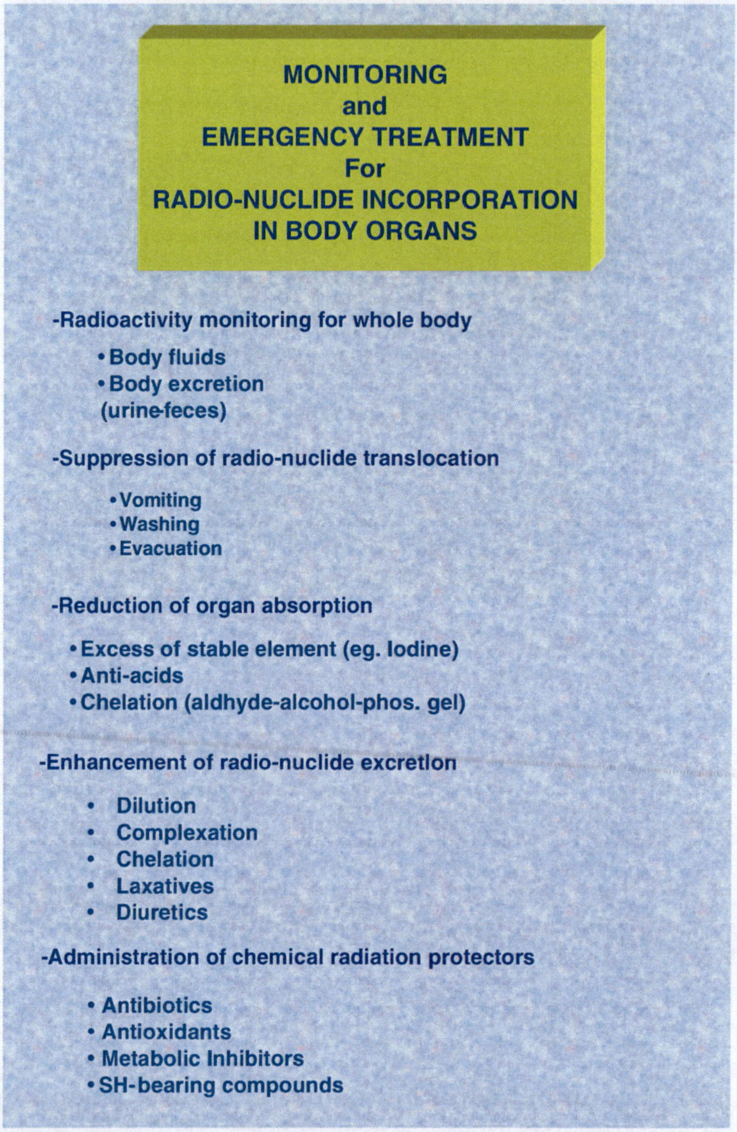

Fig. 5.15 Monitoring And Emergency Treatment For Radio-Nuclide Incorporation In Body Organs

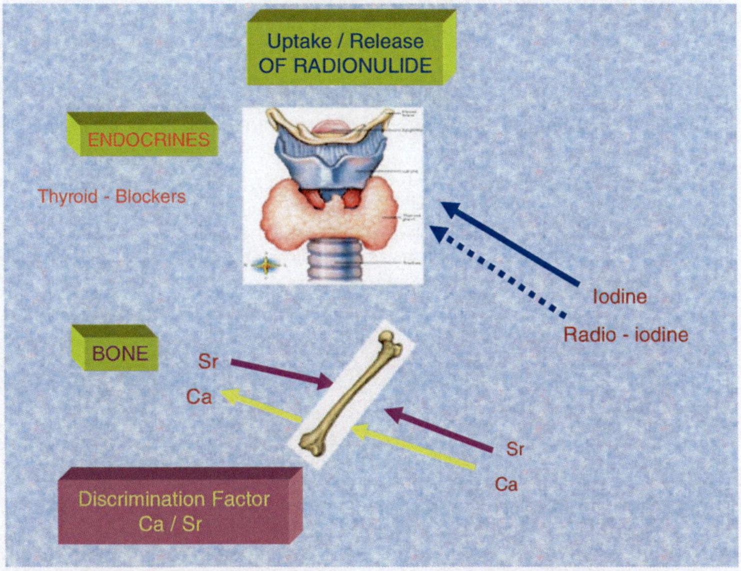

Fig. 5.16 Uptake/Release of Radionulide

5.11 Conclusion

- Environmental pollution underline potential risk on food Safety
- Radio-contamination of human food and animal feed underline substantial risk in health problems
- The synergistic effect of food radio-contamination and health problems impose aggravated health hazard
- Human food and animal feed should be routinely monitored for radioactive and chemical pollutants
- Developing Nations should regulate and enforce their own acceptable levels of radioactivity in food in view of the prevailing health conditions ,feeding habits and radiation tolerance
- Routine monitoring of radiation induced genetic and hereditary health hazard should be undertaken on human communities fed radio-contaminated food while suffering nutritional and environmental problems

Additional Readings

1. IAEA (International Atomic Energy Agency) Safety Series
2. ICRP (International Commission Radiation Protection) Publications
3. FAO (Food and Agriculture Organization) (1986) Report on recommended limits for radionuclide contamination of foods
4. Genetic and somatic UNSCEAR (UN Sci. Comm. Effect of Atomic Radiation) of ionizing radiation. UN Sci. Comm. Effect. UN Publications
5. Roushdy HM (1992) High dose radiation response at molecular and cellular levels Germany/Egypt. Course on medical aspects of nuclear and radiation accidents: Cairo, April 11–16, 1992, Julich, KFA Julich Gmbh Publications

Chapter 6
Chemical Pollutants Threatening Food Safety and Security: An Overview

Sameeh A. Mansour

Abstract Food passing several stages in the long and sophisticated food chain processing **(from farm to fork)** before being consumed and in each stage can cause morbidity and mortality, and also destruction to food industry. This is because food is a vulnerable media for contamination by thousands of biological, chemical and physical agents, and radio nuclear materials. Such contamination may occur intentionally or unintentionally far of "intended crime." On the other side, food may deliberately contaminate within "intended terroristic crime."

Food and water contamination remains the easiest way to distribute toxic agents for the purpose of terrorism. Intentional contamination of food and water supplies by such harmful agents for terrorist purpose is a real and current threat to consumer's health. This presentation addresses the chemical contamination of foods. Unlike most microbiological agents, chemical contaminants present in foods are often unaffected by thermal processing. They can be classified according to the source of contamination and the mechanism by which they enter the food product. They include a wide range of agrochemicals used in agricultural practices and animal husbandry with the intent to increase crops and reduce costs. Such agents include pesticides (e.g. insecticides, herbicides, rodenticides), plant growth regulators, veterinary drugs (e.g. nitrofuran, fluoroquinolones, malachite green, chloramphenicol), and bovine somatotropin (rBST). Moreover, food can be contaminated by chemicals that are present in the environment in which the food is grown, harvested, transported, stored, packaged, processed, and consumed. Possible contaminants include radio nuclides (e.g., cesium, strontium), heavy metals (e.g., arsenic, mercury, cadmium, copper), persistent organic pollutants, [e.g., polycyclic aromatic hydrocarbons (PAH), polychlorinated biphenyls (PCBs), dioxins, polybrominated diphenyl ethers (PBDE), acrylamide, furan, perchlorate], as well as numerous substances attributed to packaging materials [e.g., antimony, tin, lead, perfluorooctanoic acid (PFOA), semicarbazide, benzophenone, isopropylthioxanthone (ITX),

S.A. Mansour (✉)
Environmental Toxicology Research Unit (ETRU), Department of Pesticide Chemistry, National Research Centre, Tahrir Street, Dokki, Cairo, Egypt
e-mail: samansour@hotmail.com

M. Hefnawy (ed.), *Advances in Food Protection*, NATO Science for Peace and Security Series A: Chemistry and Biology 1, DOI 10.1007/978-94-007-1100-6_6,

bisphenol-A]. At present, melamine represents one of the largest deliberate food contaminant.

Increasing incidences of cancer, chronic kidney diseases, suppression of the immune system, sterility among males and females, endocrine disorders, neurological and behavioral disorders, especially among children, have been attributed to chronic exposure to toxic chemicals and/or long-term consumption of food contaminated with such chemicals. Chronic exposure to food chemical contaminants may adversely affect human health even at contamination levels below MRLs (maximum residue limits) of each chemical. Furthermore, exposure through food to different contaminants may lead to additive or synergistic effects, a matter which poses an urgent need to improve our knowledge on such possible interactions at the intestinal level.

Sophisticated analytical tools needed for detection of some chemical compounds, such as dioxins and nanoparticle materials, are not available in many countries, a matter which requires assisting these countries to raise their capacity building in this concern. On the other side, there is an urgent need for establishing a Security Plan Development (SPD) which can be realized easily and efficiently by implementing the principles of hazard analyses and critical control points (HACCP).

In conclusion, although unintentional contamination of food could be controlled by local regulations, intentional contamination for the purpose of "terrorism" should be faced internationally and early as soon as multinational consequences are expected. Coordination between food-safety authorities worldwide is needed to efficiently exchange information and to enable tracking and recalling of affected products to ensure food safety and to protect public health. Cooperation between countries has to be activated and exchanging information is something very important to minimize threat and contamination on national, regional and international levels. Prevention and response are the two major strategies for countering the threat of food terrorist, and they are everyone's responsibility all over the world in such manner.

Keywords Food safety • Food chain • Chemical contaminants • Intentional contamination • Deliberate contamination (Terrorist) • Melamine adulteration incident • Hazard analyses and critical control points (HACCP).

6.1 Introduction

Food is essential for people to live sustainable and healthy lives. Advances in food production, processing, and trade have substantially strengthened food availability, stability, access, and utilization in past decades. Yet, at the beginning of the twenty-first century, achieving food security for all is a far-reaching goal. The world has made only slow progress in reducing hunger in past decades, with dramatic differences among countries and regions. The number of undernourished people in developing countries actually increased from 823 million in 1990 to 923 million in 2007 [1]. The strong, and often interrelated, forces of change that are now rapidly transforming the world food system raise additional concerns about the future state of food security. Population and income growth—combined with high energy prices, biofuels, science and technology breakthroughs, climate change, globalization, and urbanization – are causing drastic changes in food consumption, production, and markets.

The term "food safety" refers to the extent to which food is safe to eat. The term is sometimes confused with that of "food security" which refers to the extent to which food is available – i.e. whether it is physically available and can be bought at a price that people can afford.

Food safety is a major public concern worldwide, and food consumption has been identified as the major pathway for human exposure to certain environmental contaminants, accounting for >90% of intake compared to inhalation or dermal routes of exposure [2]. About 30% of human cancers are caused by low exposure to initiating carcinogenic contaminants in the diet [3]. During the last decades, the increasing demand of food safety has stimulated research regarding the risk associated with consumption of foods contaminated by pesticides, heavy metals, and/or toxins [4].

Foodservice enterprises are unique in the business world because they accomplish the entire business cycle of making their own products, marketing those products and providing services, often under the umbrella of one single holistic organization [5].

The first security concern apparent in foodservice is the protection of employees who provide food and service [6]. However, foodservice employees are not the only persons impacted by safety and security issues. Patrons come to a foodservice with the expectation of their own personal safety [6]. Customers are the reason any foodservice exists, and it is the job of both managers and foodservice employees to see to it that customers have a positive, secure, and safe experience [7]. In addition to customers and foodservice workers, there are support personnel in foodservice such as vendors, maintenance workers, and service providers who must frequent food facilities in the performance of their jobs. These industry support personnel should be monitored for security clearance and their safety and security ensured [8].

The rapid globalization of food production and trade has increased the potential likelihood of international incidents involving contaminated food. Food safety authorities all over the world have acknowledged that ensuring food safety must not only be tackled at the national level but also through closer linkages among food safety authorities at the international level. This is important for exchanging routine information on food safety issues and to have rapid access to information in case of food safety emergencies.

The International Food Safety Authorities Network (INFOSAN) is a global network of 177 national food safety authorities, developed and managed by WHO in collaboration with the Food and Agriculture Organization of the United Nations (FAO), that:

- disseminates important global food safety information, and;
- improves national and international collaboration.

Food passing several stages in the long and sophisticated food chain processing **(from farm to fork)** before being consumed and in each stage can cause morbidity and mortality, and also destruction to food industry. This is because food is a vulnerable media for contamination by thousands of biological, chemical and physical agents, and radio nuclear materials. Such contamination may occur intentionally or unintentionally far of "intended crime." On the other side, food may deliberately contaminate within "intended terroristic crime."

The objectives of this document are: (a) to review chemicals which may contaminate food during production, processing, and handling; (b) to demonstrate health risks due to consumption of chemically contaminated foods, with special concern to chemical substances of potential hazard; and (c) to introduce some protective measures against either unintentional or intentional contamination of food, animal feed, and water with toxic chemicals.

6.2 Chemical Substances Which May Contaminate Food

6.2.1 Historical

Records of outbreaks of human illness caused by toxic contaminants in foods appeared at least several centuries BC, when mad honey poisoning was associated with an illness among troops under the command of the Greek historian and mercenary, Xenophon [9]. One of the first recorded foodborne outbreaks of ergotism occurred in Limoges, France, during the Capetian Dynasty in 994 AD [10]. The ingestion of rye bread contaminated with ergot alkaloids during this epidemic caused the deaths of approximately 20,000–50,000 victims. Numerous episodes of ergotism have occurred throughout history. Although ergotism is a possible explanation for the bizarre behavior that occurred before and during the Salem Witch Trials of 1692, there is no definite evidence that ergotism was a contributing factor [11]. In the 1950s, mining operations in the Toyama region of Japan released cadmium into the Jinzu River. The use of this water for drinking and for irrigation of nearby rice fields resulted in a disease called *itai-itai* (translation: "ouch-ouch"), manifest by osteoporosis and renal dysfunction primarily in middle-aged women [12]. In the same decade, numerous neonates born near Minamata Bay, Japan, developed birth defects and neurological abnormalities after pregnant women were exposed to seafood contaminated by methyl mercury released into the bay from a local factory. A methyl mercury-based fungicide caused an outbreak of mercury poisoning in Iraq in 1971 after grain seeds treated with the fungicide were inadvertently used for food manufacturing instead of planting [13]. Prominent outbreaks of illnesses associated with chemical contamination of cooking oils include tri-*ortho*-cresyl phosphate-induced neuropathy (Morocco – 1959), [14] yusho ("rice oil disease," Japan – 1968), yu-cheng ("oil disease," Taiwan – 1979), toxic oil syndrome (Spain – 1981), and epidemic dropsy (India – 1998). Some contaminants are unavoidable in food manufacturing. Food products are considered *adulterated* when concentrations of avoidable contaminants (e.g., pesticides, melamine) exceed established standards, sometimes prompting food recalls after sale and distribution [15].

6.2.2 Contaminant Pathway (From Food to Human Body)

The chemical industry has enjoyed spectacular growth in the last century. There are now more than 100,000 different chemicals available on the market. Chemicals are incorporated into countless consumer products, some of which undoubtedly benefit our standard of living. But they also provide a source of daily exposure to a cocktail of hazardous chemicals. Undoubtedly there are many routes of exposure which contribute to the patterns of contamination observed. Food has long been thought to be the primary route of exposure for most persistent and bioaccumulative chemicals. However, in recent years greater attention has been given to the potential exposure directly through the use of products containing hazardous ingredients and indirectly through their contamination of the indoor environment [16].

Most persistent and bioaccumulative chemicals eventually find their way into our bodies via the food chain. They are lipophilic (fat soluble) and therefore tend to build up in the fatty of animals, a process known as bioconcentration. As one animal consumes another in the food chain, the levels may become even higher (biomagnification). Consequently, predatory animals at the top of food chains tend to accumulate the highest levels of some of these hazardous chemicals. Humans too are vulnerable, as our diet includes other animals. The magnitude of biomagnifications for PCBs was illustrated by a study of birds' eggs from Lake Ontario in the United States. The concentration of PCBs in the eggs was 25 million times higher than that in the lake water, having accumulated through the food chain to the fish on which the birds were feeding [17].

Hazardous chemicals can end up in the bodies of unborn children via the mother. Research shows that mothers pass some chemicals to their child both during pregnancy and during breastfeeding. Persistent lipophilic chemicals that have accumulated in the bodies of women over their lifetime are released as fat reserves are mobilised, for instance when they are pregnant or breast feed their children. From studies carried out on animals it appears that phthalates, for example, can pass across the placenta [18].

The hormone-disrupting effects of hazardous chemicals remain a potentially severe threat. In recent decades, the concern about chemicals with hormone disrupting properties has escalated, as several wildlife species have certainly been affected by exposure to endocrine disrupting chemicals (EDCs). The possible effects of chemical substances with so-called 'oestrogenic' or 'anti-androgenic' activity on the hormone system has been of particular concern in the last decade. These chemical substances mimic the effects of female sex hormones (oestrogens) or block the action of male sex hormones (androgens) and can seriously disturb reproduction in animals. For example, if male embryos of rats or mice are exposed to chemicals such as bisphenol-A, DDT or vinclozilin (a fungicide), such exposure can negatively affect the development of male sex organs [19].

6.2.3 How Are Humans and Wildlife Exposed?

Chemicals contaminants may reach the environment as a result of direct discharges from industrial processes, inappropriate disposal of waste, and "leaching" from waste and landfill sites. Direct application (e.g. old organochlorine pesticides) and spillages or leakages during manufacture, transport, or storage are also important inputs to the environment. Chemicals can also escape into the environment during their incorporation into consumer products or from products as they are used and wear out. For example, the air and dust in our homes, schools, and offices can contain chemicals such as flame retardants, phthalates, and PCBs that have escaped from furniture, flooring, building materials, and electrical appliances.

For some chemicals used in consumer products, the route of human exposure is relatively direct e.g. inhalation [20] of flame retardants in indoor air and dust [21]. Chemicals found in personal care products such as toiletries and cosmetics can also enter through the skin e.g. phthalates and synthetic masks [22]. However, for humans and wildlife alike, the most important exposure route for many of the chemicals discussed here, particularly those which are persistent and bioaccumulative such as DDT and PCBs, is the diet. This is the result of contamination of the food chain, and the presence of man-made chemicals in food is illustrative of the truly global scale of the chemical contamination problem. Food is a crucial link in a chain of events that begins with the manufacture of chemicals and ends with their unwelcome appearance in the blood supply of developing fetuses and the blood and tissues of children, adults, and wildlife species.

Organic chemical hazards to foodstuffs have been discussed in detail elsewhere (e.g., [23, 24]). Although there are multiple exposure routes into foodstuffs for most chemicals, there are basically four main classes of organic chemical hazards that should be considered, i.e., residual chemicals, applied chemicals, accidental chemicals, and background chemicals [25]:

1. *Residual chemicals*: these are contaminants already present within or on a foodstuff resulting from previous exposure of ingredients or raw materials during food production, processing, or packaging. Ideally, all foodstuffs should be supplied with documentation identifying potential residues present and all activities, procedures, and testing conducted to minimize any associated hazards. Current attention in chemical HACCP (Hazard Analysis and Critical Control Point) is mainly focused on residual chemicals from the agricultural sector (e.g., pesticides, growth hormones, fumigants, and some natural toxins).
2. *Applied chemicals*: these are chemicals that are applied intentionally to foodstuffs, and intended to be retained, e.g., food additives and food preservatives.

Very few applied chemicals are likely to be used in the producer-to-consumer sector, with the possible exception of food protection chemicals, such as pesticides and fumigants, applied during either transportation or storage operation to prevent food infestation.

3. *Accidental chemicals*: these are applied accidentally or generated unintentionally. Examples include impurities in applied chemicals and migratory species from materials such as packaging. Foodstuffs can also be exposed to many other site chemicals as the result of inappropriate handling practices or local accidents (e.g., spillage or site fire). Relevant site chemicals are likely to include cleaning materials, disinfectants, machine lubricants, and paints. With regards to generated accidental chemicals, most 'off-the-shelf' foodstuffs do not undergo processing in the producer-to-consumer sector and attention is normally focused on chemicals generated during storage, e.g., toxic products of fermentation or oxidation processes (e.g., [26, 27]). However, in cases where food preparation is conducted within the producer-to-consumer sector (e.g., in on-site bakeries, restaurants, and cafés), chemicals generated during both processing and storage should be considered.
4. *Background chemicals*: these are ubiquitous environmental contaminants that may enter the food chain at almost any stage of food production, distribution, and handling. Research was initially focused on a small number of contaminants, i.e., polycyclic aromatic hydrocarbons and polychlorinated biphenyls, dibenzo-p-dioxins and dibenzofurans. However, other classes of contaminants are now receiving similar attention, e.g., volatile aromatics, chlorinated solvents, benzenes, naphthalenes and diphenoquinones, polychlorinated diphenyl ethers, polybrominated dioxins, biphenyls and diphenyl ethers, and synthetic musks. Most sites used in the producer-to-consumer sector are likely to be purpose built or specially modified enclosed sites (e.g., warehouses, storage sites, and retail outlets). Furthermore, working practices at these sites are likely to be regulated and the associated workforces are required to conduct all their activities in accordance with sanitary working practices. Consequently, many of the exposure routes commonly associated with environmental contamination (e.g., soil, river water, rain water) are likely to be insignificant. The main background chemical exposure routes to foodstuffs in such sites are, therefore, likely to be local air, site water supply, and contact transfer (e.g., from handling equipment, site equipment, members of the site work force or, in the case of retail outlet, potential consumers).

6.2.4 Fear of Food Contamination

Bánáti [28] in his article demonstrated fear of foods through Hungarian experience. He reported that 42% of Europeans believe that the food they eat is harmful for their health [29]. There have been several food scandals and food-related outbreaks in the last decade all over Europe. Besides the widely known bacterial infections occurring from time to time not causing any unexpected uncertainties, certain psychrotrophic pathogens such as Listeria and other emerging pathogens, i.e. prions causing BSE (the so-called mad cow disease) or the highly pathogenic H5N1 virus being responsible for avian flu, gained attention of consumers throughout Europe.

The emerging risks, the fear of the unknown, especially of new techniques and technologies (e.g. modern biotechnology) attacking the genome thought to be intact or factors being considered as "unnatural" or artificial (e.g. hormones in meat and/or cloning animals for food) make consumers worried despite the safety of their food declared by experts. Despite the fact that food has never been safer, it seems that consumers are considerably uncertain, anxious, and increasingly critical about the safety of their food.

Forty-two percent of Europeans consider it likely that the food they eat will damage their health. No single problem or risk stands out as being the culprit. Besides food poisoning (16%), chemicals, pesticides and toxic substances (14%), obesity and overweight (13%) are listed as possible problems or risks associated with food according to a recent Europe-wide study [29]. When respondents were confronted with a list of possible problems, 61% were worried or very worried about bacteria like Salmonella and Listeria, 63% about pesticide residues, and almost the same number of consumers (62%) was afraid of antibiotics residues and hormones, and also of new viruses like avian influenza (62%). More than half of the consumers (53%) are worried about BSE. Even more of the European respondents have chosen additives (57%), GMOs (58%) and pollutants such as mercury and dioxins (59%) from the list of potential risk factors related to foods [28]. This fear could be enhanced by inaccurate or incorrect information provided either by relevant authorities competing in a fragmented system of official control to exert power or by the media spreading information not based on sound science or willing to cause sensation and gain attention.

The application of nanotechnology in the food sector, such as GMOs, could be the next new technology not to be accepted and feared, if no societal dialogue is started to inform the society based on carefully chosen and targeted information and properly communicated potential risks and benefits [30].

6.2.5 Is Organic Fruits Are Free of Contaminants (e.g., Pesticides and Heavy Metals)?

During the last decades, the increasing demand of food safety has stimulated research regarding the risk associated with consumption of foods contaminated by pesticides, heavy metals, and/or toxins [4]. Chemophobia is the most common reason for the public to choose organic food on the assumption that such food is free of synthetic pesticides. Organic farms distinguish from all other forms of farming by a rejection of soluble minerals as fertilizers and synthetic pesticides in favor of natural ones [31]. Subsequently, this has resulted in increasing demand for organic produce of different varieties of vegetables, fruits, milk products, and cereals in many parts of the world.

The literature offers much evidences regarding occurrence of contaminants (e.g., pesticides and heavy metals) in the organically farmed crops [31–35]; even though at levels below the allowed by Codex Alimentarius. In organic farming, the producer attempts to reduce quantities of pesticides and fertilizers applied to their fields. This may interpret the lower levels of pesticides and heavy metals in organically-farmed crops compared to the conventionally-farmed ones. According to recent studies, contamination of cucumber fruits produced from three different types of farming (e.g., conventional, greenhouse and organic), either by pesticides or heavy metals, was varied from a season to another. Generally, the greenhouse cucumber was found to contain the highest value of total pesticide residues (1.016 mg kg^{-1}), followed by organic (0.442 mg kg^{-1}) and then conventional (0.415 mg kg^{-1}) cucumbers. Heavy metal contamination in the three cucumber types accounted to 4.968, 5.350 and 6.248 mg kg^{-1}, respectively [36].

Contamination of conventionally-farmed potatoes was found two times, nearly, that of organically-farmed potatoes either by pesticides or heavy metals [37].

Both studies shed light to the problem of multi toxicants in food commodities such as cucumber and potatoes, common elements in the daily human diet.

6.3 Contamination of the Food Chain

Food represents a part of the global environment which can be contaminated by chemicals from many different sources. Following their release into the environment (soil, sediment, water, air) chemical contaminants can enter plants and animals at the bottom of the food chain which are then consumed by animals higher up. The chemicals contained in these animals and plants can enter our own bodies when we consume them as food e.g. meat, dairy products, fish, vegetables, fruits. This "food chain" route of contamination is especially important for chemicals that persist and accumulate in the environment, such as DDT, PCBs and brominated flame retardants [38]. It is also relevant for chemicals that are used in large amounts and occur ubiquitously in the environment e.g. phthalates. Packaging and processing may also introduce chemicals into food e.g. perfluorinated chemicals used in greaseproof packaging for fast foods. It is for these reasons that WWF's focus now shifts to investigating contaminants in food, and in the process presents the next logical step on from its human biomonitoring work [39].

For the purpose of this chapter, we will consider that contamination of food may occur unintentionally by environmental chemicals, such as those mentioned above under the class of "background chemicals." But food may intentionally contaminate through two completely different purposes: (a) unintended harm; (b) intended harm. In the first route, contamination occurs due to applying or adding chemicals to improve or secure food quality and quantity, within "Good Agricultural Practices (GAP) or Good Manufacturing Practices (GMP)". For example, applying pesticides

Table 6.1 General overview of the typical food-chain and types of major chemicals which may contaminate our food

Stage	Chemical contaminant
I. Agricultural (and Farm) production	Pesticides, heavy metals, veterinary drugs, hormones, antibiotics, POPs, natural toxins
II. Storage and transportation of raw commodities	Pesticides, cleaners, POPs, natural toxins
III. Processing and manufacture	PCBs, furans, food (and Feed) additives, heavy metals, acrylamide, cleaners, packaging materials
IV. Storage and transportation of processed and manufactured products	Pesticides, cleaners, POPs, natural toxins
V. Wholesale and retail distribution	Pesticides, cleaners
VI. Food service sector	Pesticides, cleaners

on crops to combat pests attacking crops in the fields and stores, or adding additives to improve or achieve amiable taste, color, or flavor for food. Such agents may pose a risk of harm when their concentrations exceed established standards. The second route of contamination goes through well-planned actions for the purpose of inducing mass harm to human health; that is what so-called "food terrorism."

Therefore food, during its production chain, may contaminate with different types of chemicals such as: Chemical warfare agents (CWA's); Pesticides; Heavy metals; Persistent Organic Pollutants (POPs); Food additives; Feed additives; Veterinary medicines; Hormones; Packaging – released chemicals; Cleaners; Chemicals of natural origin; and many Others. Table 6.1 demonstrates types of chemicals with respect to the different stages of food chain.

The "Chain of Contamination" as demonstrated by the WWF means that the global food chain is contaminated with hazardous chemicals such as DDT, PCBs, and brominated flame retardants. As a result, food is a major source of exposure for humans and wildlife. Meat, fish, dairy products, vegetables, and fruits can all contain traces of man-made chemicals [39]. Hazardous chemicals are commonly found in human blood [40] and can be passed to a developing baby through the placenta [41] and breast milk [42]. The main routes by which humans are exposed to chemicals are through food, inhalation of indoor (and outdoor) air, and absorption through the skin. Chemicals can escape from consumer products during use and leach out of landfills following their disposal. They can enter indoor air and dust [21, 43] and reach soil, groundwater, rivers, and oceans – exposing wildlife and humans. Persistent man-made chemicals can increase in concentration as they move up the food chain for example, from plankton to fish to seals to polar bears accumulating in the bodies of wildlife species and contributing to serious health impacts [44]. Hazardous chemicals can be released into the environment during manufacture, transport, storage, and the disposal of waste, as well as directly from consumer products. They can travel via air and water currents around the globe and accumulate in soils, sediments, rivers, estuaries, and oceans and enter the food chain of wildlife and humans. Even remote, pristine areas such as the Arctic, far from industrial activities, are polluted with manmade chemicals [45]. Currently only 14% of the chemicals used in the largest volumes have the minimum amount of data publicly available to make an initial basic safety assessment [46]. Some of these chemicals have hazardous properties – they may disrupt hormones, or be carcinogenic, persistent, bioaccumulative, or toxic to reproduction.

6.4 Chemicals of Special Concern as Food Contaminants

A brief account on some chemicals of special concern to food contamination is given below.

6.4.1 Persistent Organic Pollutants (POPs)

Persistent organic pollutants (POPs) are organic (carbon-based) compounds that include synthesized substances of agricultural and industrial usages (e.g. pesticides and polychlorinated biphenyls; PCBs). Other substances are by-products generated as a result of human and natural activity (e.g. dibenzo-p-dioxins 'dioxins' and dibenzo-p-furans 'furans') of which human activity accounts for the major proportion of releases [47]. POPs are toxic compounds, causing adverse health effects, such as birth defects, damage to immune and respiratory systems, and critical organs. Hormone system dysfunction associated with POPs includes damage to the reproductive system, sex-linked disorders, and shortened lactation periods for nursing mothers. As well, endocrine disruption can have developmental and carcinogenic effects. Moreover, exposure to POPs can result in death in humans (including aborted fetuses) and wildlife. Exposure of the fetus to minute concentrations of some POPs can result in neurophysiological effects, such as attention deficits, learning disorders, behavioral problems (e.g. increased aggressivity), and poor gross and fine motor coordination [48].

Man-made chemicals are an integral and vital part of our modern lifestyles. They are found in a vast range of consumer products – from furniture, clothing, and toiletries to electrical appliances, car interiors, and cleaning products. While they have undoubtedly improved the quality of our lives, many possess undesirable properties. They can be harmful to health, and many can persist in the environment and Bioaccumulate in the bodies of wildlife and people. These properties have resulted in ecosystems all over the world being contaminated with a cocktail of man-made chemicals. Examples include the chemicals DDT (an insecticide) and PCBs (polychlorinated biphenyls – used in electrical components), which despite having been banned for decades, are still found throughout the global environment.

In more recent years, modern chemical compounds such as brominated flame retardants (used to prevent fire in plastics e.g. TVs, computers and textiles e.g. furniture, carpets) and perfluorinated "non-stick" chemicals, (used for waterproof and stainproof coatings) have followed PCBs and DDT to all corners of the globe. Some chemicals can also interfere with hormone processes in the body – these are known as "endocrine disrupting" chemicals (EDCs). Examples include phthalates, primarily used to soften plastics and found in numerous consumer products, from vinyl flooring to cosmetics [39].

There is a large body of scientific evidence on the adverse impacts of man-made chemicals on wildlife species e.g. population crashes in birds of prey caused by DDT, immune impacts of PCBs on seals. Research, including WWF's own biomonitoring studies [40, 49] has also consistently shown that humans all over the globe are exposed to a cocktail of potentially hazardous chemicals including DDT and PCBs, as well as brominated flame retardants, perfluorinated chemicals, artificial masks (used as synthetic fragrances in many consumer products) and phthalates. Independent researchers also investigate contaminants in food. For example, recent studies have investigated brominated flame retardants [38] and nonylphenols [50] in food. WWF is of the opinion that chronic, low level exposure to a combination of chemical contaminants via the diet and other exposure routes has not been given sufficient consideration in past decision making on chemicals. WWF is therefore lobbying to ensure the EU's proposed REACH legislation is sustainably strengthened, as it offers a once in a lifetime opportunity to tackle the problem of food chain contamination by driving the substitution of persistent, bioaccumulative and endocrine disrupting chemicals with safer alternatives.

Many of these chemicals have been detected in young children as well as adults, and in some cases at higher levels in children than in adults. Alongside this, there is growing concern over possible links between certain chemicals (particularly endocrine disrupting chemicals) and human health impacts such as cancer, reproductive problems, birth defects, asthma, allergies, behavioural problems, disruption of infant brain development, cardiovascular disease, diabetes, and obesity [39, 48].

6.4.1.1 Organochlorine Pesticides

Most of the organochlorine pesticides (OCPs) have been banned since decades in different parts of the world including Egypt, but their residues still appear as pollutants in food as well as in the environment. Their occurrence and long-range transport at local, regional, and global scales has been recently explored [51]. Frequent detection of these compounds in different environmental compartments could provide information about to what extend the threats posed by these xenobiotics may exist.

Hydrophobic OCP uptake by different vegetable crops including cucumber, potatoes, carrots, tomatoes, leeks, cabbages, spinaches, and others had been widely reported in the literature (e.g., [35, 52, 53]). In these studies, crops were grown in soils treated with the insecticides up to 5 and 20 years earlier. Vegetables grown in conventional or organic soils have shown to accumulate OCPs efficiently with residue levels 4- to 45-fold greater than those in soils [33]. In addition to plant uptake, the occurrence of OCPs may refer to atmospheric deposition of these volatile and semi-volatile compounds on the cultivated crops [31, 32].

Well-known compounds such as DDT and its metabolites belong to the group of organochlorine pesticides. These chemicals were widely used in the past all over the world. Although their manufacture and application are now largely prohibited or restricted in industrialized western countries, they can still be found in the environment, in wildlife and in humans due to their persistence. As DDT bioaccumulates in animal fat, most humans are exposed to this substance primarily through their food. Residues have been found in human blood, serum, and breast milk. Reproductive disorders are well documented in animal studies, and exposure is also linked to human developmental disorders. The main DDT component, p,p'-DDT, is classified as possibly carcinogenic to humans [54].

As in other studies, HCB, p,p'-DDE and p,p'-DDT were found in the majority of samples (75% or more), in maternal as well as cord blood. Although this frequency is comparable to other studies, the concentrations of p,p'-DDE found in maternal blood in the current study (0.33–1.9 ng/g serum) are somewhat lower than those reported in previous studies [55, 56]. However, the concentrations of p,p'-DDT (0.09–1.5 ng/g serum) were similar to or slightly higher than those in the WWF study [56].

Other chemicals have also been associated with effects on brain function and with thyroid disruption. Some pesticides, especially organophosphates, DDT, pyrethroids and paraquat, are particularly under the spotlight with regard to neurotoxic effects [57–59].

6.4.1.2 Polychlorinated Biphenyls and Dioxins

PCBs and dioxins may not be the only man-made chemicals implicated in affecting the brain development of our children. Some researchers suspect that certain brominated flame retardants, which are used in many electrical and electronic appliances, disrupt brain development early in life. The structure of these chemicals, termed PBDEs, is remarkably similar to PCBs. PCBs have been found to cause immune suppression, altered sexual development, cancer, delayed brain development and, lower IQ and were banned in 1976. It has also been suggested that they are linked to behavioral problems like hyperactivity in humans. As with PCBs, exposure to PBDEs may be particularly harmful during a critical window of brain development during pregnancy and early childhood [60].

The biomonitoring study recently conducted by WWF, involving 155 people from the UK, found that women commonly had lower amounts of PCBs in their bodies than men. The more children the woman had given birth to, the lower the PCB concentration [61]. This suggests that a significant proportion of the body burden of persistent chemicals accumulated by the mothers may have been lost during pregnancy, including a proportion which would inevitably have passed to the developing child.

6.4.1.3 Brominated Flame Retardants

PBDEs may have a similar range of exposure sources as the phthalates. Food is almost certainly the main source of exposure to some of the more bioaccumulative (lower brominated) PBDEs. PBDEs have been detected in fish and shellfish in the range of 21–1,650 pg/g fresh weight [62]. In comparison, beef, pork, and chicken contained 6.25–63.6 pg/g, and three different vegetables had levels of 38.4–134 pg/g. Ohta et al. [62] showed a strong correlation between PBDE levels in breast milk and intake of fish and shellfish. For chemicals such as decabromodiphenyl ether (BDE-209 or 'deca'), however, more direct exposure to e.g. contaminated indoor dusts or even direct contact with products may be relatively more significant in terms of human exposure. PDBEs have been found in human breast milk, blood, and adipose tissue (e.g., [63–65]).

Extensive breast milk studies in Sweden show an exponential increase in PBDEs in breast milk (an average increase from 0.07 to 4.02 ng/g lipid between 1972 and 1997) [66]. However, a recent paper has reported a decrease of PBDEs in Swedish breast milk since 1997, possibly due to a voluntary phase out of penta-BDE [67]. Babies born to mothers in the USA may be more at risk of PBDE contamination than in Sweden and Norway. Mazdai et al. [68] found that the concentrations of PBDEs in maternal and foetal serum samples in Indianapolis, USA, were 20–106 times higher than the levels reported previously in Swedish mothers and infants [63] and 20 times higher than Norwegian blood samples [69].

A range of PBDEs and brominated bisphenol-A compounds, such as TBBP-A, show oestrogenicity in human cells lines and bind to the oestrogen receptors [70]. The metabolism of PBDEs to hydroxylated-PBDEs produces more potent oestrogen mimics. Brominated bisphenol A compounds with the lowest bromination showed the highest effect, and among the PBDEs, BDE-100, BDE-75 and BDE-51 showed the highest activity [71]. Toxicity studies of BFRs in animals are limited and primarily consist of high dose studies of PBDEs. Nevertheless, there is evidence that chronic exposure to PBDEs can cause birth defects in rodents [72, 73], as well as impacts on nervous system and behavioural development.

6.4.1.4 Phthalate Compounds

Phthalate exposure is widespread and continuous, a result of their high volume use in PVC and other open applications, which has led to them becoming one of the most ubiquitous man-made chemicals in our environment. Our exposure to them can result from leaching from products such as soft PVC (vinyl) flooring, furnishings, clothing etc. as well as through inhalation of contaminated indoor air, exposure to household and/or office dusts, consumption of contaminated food or, in some cases, of contaminated drinking water. Concentrations in house dust can reach several milligrammes per gramme of dust (parts per thousand). The use of phthalate esters in products such as perfumes, which may contain high levels in particular of the phthalate DEP as a solvent and alcohol denaturant, may result in additional exposure. With respect to food intake, presence of phthalates in food contact materials may be of particular concern [22].

To date, very few studies have focused directly on possible impacts of endocrine disrupting chemicals on human male reproductive health. Two of the most recent and most prominent of such studies [74, 75] relate to exposure in the womb and shortly after birth to phthalates (phthalate esters).Taken together, these studies suggest that exposure to phthalates in the early stages of life is associated with hormone disruption and impacts on male reproductive health.

Phthalates have also been found in breast milk [76, 77]. Additional concerns were raised by a study of Swedish breast milk in which concentrations of the brominated flame retardant pentabromodiphenyl ether (penta-BDE) doubled every 5 years over a period of 25 years from 1972 [66]. More recently, the concentrations have declined as a result of tighter controls on their use. Synthetic musk compounds [78] and nonylphenols also appear in breast milk [50].

Metabolites of phthalates are detectable in urine samples from adults indicating exposure to phthalates [79–81]. Animal studies show that phthalates cross the placenta and pass into breast milk [18, 76, 77]. Their ability to cross the human placenta and therefore reach the developing child in the womb was confirmed by the common occurrence of phthalate esters, especially DEHP, DBP, BBP and DEP, in samples of umbilical cord blood [16].

Children may be more exposed to phthalates than adults. For example, in one study, of the seven urine phthalate metabolites tested, the highest levels of metabolites for DEHP, DBP and monobenzylphthalates were found in the youngest age group tested: the 6–11-year-old children [80].

6.4.1.5 Alkylphenols

Despite increasing restrictions on their use in Europe, alkylphenols remain widespread as contaminants in our environment, including in our food. For example, a study of 60 food-products on the market in Germany illustrated the widespread nature of alkylphenol contamination [50]. These were all popular, common foods in Germany, including 39 adult foods, from marmalade to sausages, 20 baby foods, and 1 sample of breast milk. Nonylphenol was detected in every sample within the range of 0.1–19.4 μg/kg, and was not related to fat content. The authors stressed that, since the foods varied substantially in nature, source, preparation methods and packaging, it is likely that there are multiple entry points for nonylphenol into the human food supply.

There have been few studies on levels of human contamination by alkylphenols, but those that have been performed clearly show that children are contaminated before birth (umbilical cord) and after birth (breast milk). Nonylphenol has been detected in human umbilical cords [82], confirming that it crosses the placenta from the contaminated mother to the growing foetus.This was more recently reaffirmed by the joint Greenpeace/WWF study into chemical contaminants in human umbilical cord blood donated by volunteer new mothers in the Netherlands [16]. Nonylphenol itself was detected in 12 of the 17 cord blood samples analysed in this study.

6.4.1.6 Artificial Musks (Nitromusks and Polycyclic Musks)

Although historically nitromusks (including musk xylene, MX and musk ketone, MK) dominated the European market for fragrance additives, their place has since been taken by the polycyclic musks, especially AHTN and HHCB. Their use in cleaning and personal care products is thought to remain widespread, though few product-specific data exist.

Greenpeace published in February 2005 a report quantifying the presence of a range of synthetic musk compounds, including AHTN and HHCB, in perfumes sold in Europe [22]. Nitromusks have been found in adult human adipose tissue and blood [83, 84]. Several studies have reported the presence of these chemicals in breast milk. In addition, both HHCB and AHTN were frequently found in samples of human umbilical cord blood in the Netherlands [16], albeit at low ppb levels in the serum, with HHCB detected in all but 1 of the 27 samples analysed.

MX and MK possess oestrogenic activity in vitro with MK showing an affinity for the oestrogen receptor three times greater than MX [85]. However, when MK is reduced to its metabolite, it loses its activity, whereas when MX is converted to p-amino-musk xylene, its oestrogenic potency increases [85]. The polycyclic musks, AHTN and HHCB, induce both oestrogenic and anti-oestrogenic activity depending on the cell type and the receptor subtype targeted. Weak oestrogenic effects are observed at relatively high concentrations (10 μM) while anti-oestrogenic effects are seen at 0.1 micromolar [86], including effects in whole organisms.

6.4.2 Organotin Compounds

Organotins are hormone disrupters probably best known because of the devastating effects of tributyl tin (TBT) used in antifouling paints on certain marine molluscs. However, both TBT and the pesticide triphenyl tin (TPT) have been observed also to inhibit a variety of enzymes responsible for the production of male and female sex steroid hormones in other organisms, including testosterone and oestradiol [87–89]. While low-dose developmental studies are lacking in mammals, insufficient activation of male hormones is known to be responsible for developmental disorders of the male reproductive tract. A recent study suggested that organotins may also cause developmental effects in utero at relatively low doses by targeting the maternal thyroid [90]. Effects varied depending on dosage but appeared to be linked to the reduction of maternal serum thyroxine and triiodothyronine throughout gestation. Effects included reduced maternal weight gain, increased post-implantation loss, decreased litter sizes, decreased foetal weights, delayed foetal skeletal development, and abnormalities in genital development in male fetuses.

The presence of TBT in seafood, primarily a result of its former use as an antifouling agent in ship hull paints, has led, in some regions, to elevated intakes. In Japan, for example, where fish is a major part of the diet, the estimated daily intake of the organotin TBT is relatively high (2.5 μg/kg body weight based on a fish consumption of 150 g/day) [91]. At the same time, we may be exposed to TBT and other organotin compounds, including the mono- and dibutyl forms (MBT and DBT) used inter alia as stabilizing additives in PVC, from a range of other, sometimes rather unexpected, sources. For example, TBT, DBT and MBT have all been reported to leach from some brands of baking parchment, and DBT and MBT from gloves for kitchen work, dish-washing sponges, and cellophane film for food, on sale in Japan [92]. Although organotins, particularly TBT, have been reported in a wide range of molluscs, fish, marine birds, marine mammals, and freshwater birds [93], aside from a few reports, levels of contamination in humans are largely unknown, and there are no readily available reports on child contamination.

6.4.3 *Heavy Metals*

It has been long recognized that heavy metals are among the major contaminants of food supply and may be considered as the most important problem to our environment [94]. Heavy metals such as Cd and Pb have been shown to have carcinogenic effects [95], and high concentrations of Cu, Cd, and Pb in fruits and vegetables were related to high prevalence of upper gastrointerestinal cancer [96]. Heavy metal contamination may be occurred due to irrigation with contaminated water, the addition of fertilizers and metal-based pesticides, industrial emissions, transportation, harvesting process, storage, and/or sale. It is well known that plants take up metals by absorbing them from contaminated soils as well as from deposits on parts of the plants exposed to the air from polluted environments [97].

Metal contaminants such as lead, mercury, arsenic, and cadmium come from factory emissions, mining operations, and metal-containing industrial products used in food production. Methyl mercury found in commercially sold seafood is deemed an unavoidable contaminant because contamination pre-exists in the raw material; therefore, the contamination does not result from food processing or distribution. Fish and shellfish acquire methyl mercury primarily from micro-organisms that methylate environmental inorganic mercury compounds released primarily from industrial sources [98].

Metals have preferential accumulation in human tissues. For example, arsenic, chromium, cadmium, lead affect the nervous system and kidneys, as well as hematopoietic system (blood). Cadmium and chromium affect liver, and in addition to lead, they harm the reproductive system in male [99].

6.4.4 *Bisphenol-A*

Many tin can linings, clear plastic re-usable water containers, baby feeding bottles, and white dental fillings are made from polymers that can release bisphenol-A (BPA) or related compounds during use. BPA is a widely used intermediate in the production of epoxy resins, polycarbonate plastics, and flame retardants. Bisphenol-A is the building block or monomer used to form polycarbonate, a plastic which is used in situations where it can come into contact with food. Non-polymerised BPA may be released from the polycarbonate. BPA was found in canned food [100], due to migration of material on the inside of the can to the food. BPA was found in about 40% of the blood samples in the previous Greenpeace study [101]. A study of BPA in the blood of pregnant women and in placental tissue and umbilical cord blood has shown that exposure levels to BPA were similar to those suggested to be toxic to reproductive organs of male and female offspring in animal studies [102]. In another study, the chemical was also reported in maternal serum at concentrations of 0.21–0.79 ng/g and in cord blood serum at 0.45–0.76 ng/g [103].

Bisphenol-A, is known to have oestrogen (female sex hormone)-mimicking properties. It is reported to cause sex changes [104], reduced nursing behaviour [105], more masculine play behaviour in females [106] and increased aggression in male animals [107]. Furthermore, bisphenol-A has been found to abolish the sex differences in open-field behaviour [108]. It binds to oestrogen receptors range of human cell lines and mimics all the oestrogenicity parameters, confirming it as one of the stronger oestrogenic chemicals [70, 71]. Scientists also suspect that man-made chemicals may be contributing to a range of learning disabilities, including attention deficit hyperactivity disorders (ADHD), which are becoming increasingly prevalent in children. Although many factors are liable to be implicated in causing ADHD, neurotoxic chemicals may be contributing to its incidence [109].

Nonylphenols and bisphenol-A have both been found in the umbilical cords of newborn babies [82, 102] and in the amniotic fluid [110], suggesting that these chemicals can pass across the placenta.

The health effects of bisphenol-A have been demonstrated in an ever- increasing number of animals studies at levels up to 2,500 times lower than the EPA's 'lowest observed dose effect' dose. Induction of early puberty in laboratory animals can occur at extremely low doses [111]. Human infants ingest bisphenol-A in formula at an estimated daily rate of 1.6 μg/kg/day, giving little safety margin for doses that cause effects in animals (as little as 2 μg/kg/day) [112].

While there are many potential exposure routes, the principal concern regarding exposure to bisphenol-A and its derivatives remains contamination of food. Bisphenol-A can leach into food as a result of the presence in food contact materials of un-reacted compound as well as from degradation of polymers such as polycarbonate (typical plastic used to make baby bottles). It has been found to migrate from rubber products and plastic stretch-film used in food-contact applications [113, 114], as well as from the lining of many food cans. Bisphenol-A levels migrating from nine dishwashing, boiling, and brushing [115]. In some instances, contamination has even been reported to arise from water filters [116]. Furthermore, patients on kidney dialysis may receive elevated exposures as a result of the use of polycarbonate components in the equipment [117].

The subject of the effects of bisphenol-A on humans has been contentious. It has been suggested that it is only partially absorbed, has a high conversion rate to the biologically inactive bisphenol-A monoglucuronide, is rapidly excreted, and shows no evidence of bioaccumulation in tissues [102]. For these reasons, until recently, many scientists believed that the active parent form of bisphenol-A would not be found in human plasma, and therefore no significant levels could reach the fetus. However, studies from Germany and Japan have now confirmed that children are exposed to bisphenol-A before birth. The first, a Japanese study, found bisphenol – in umbilical cords [82]. Studies on mice and monkeys then showed that this chemical can cross the placenta [118].

Data from Ikezuki et al. [110] suggest that bisphenol-A may concentrate in amniotic fluid as it was found at approximately 5-fold higher concentrations at 15–18 weeks gestation, compared with other fluids. Schonfelder et al. [102] also showed that in 14 of 37 cases, foetal plasma levels of bisphenol-A were higher than in the corresponding mother's blood. Foetal levels of bisphenol-A were also significantly higher in males, which may indicate sex differences in the metabolism of this chemical. Takeuchi and Tsutsumi [119] also found this gender difference in a study on adults, and suggested that it may be due to androgen (male hormones) related metabolism of bisphenol-A.

6.4.5 Cyanide

Hydrogen cyanide is a colourless gas or liquid and has an odour of bitter almonds. It is widely available and its industrial use includes precious metal extraction and electroplating. It is highly volatile. Although this high volatility limits its use outdoors, it can be highly lethal in confined spaces.

Cyanide is normally detoxified by the enzyme rhodanese (thiosulfate–cyanide sulphur transferase), which binds sulphur to cyanide forming thiocyanate, which is excreted in urine. This detoxification is easily overwhelmed and is only protective against low-level exposures. The US government considers cyanide to be among the most likely agents of chemical terrorism. Cyanide differs from many other biological or chemical agents for which little or no defense is available because its individual and public health effects are largely remediable through appropriate preparedness and response [120]. Cyanide's toxic mechanism is interruption of the citric acid cycle by reversibly binding with cytochrome oxidase and halting oxidative phosphorylation. This ends aerobic energy production and rapidly results in cell death. Symptoms of mild poisoning include headache, dizziness, drowsiness, nausea and vomiting, and mucosal irritation. Severe effects include dyspnoea, impaired consciousness and coma, convulsions, tachyand bradydysrhythmias, hypotension, cardiovascular collapse, and death. Death can occur within minutes of inhalation. Metabolic acidosis results from lactic acid accumulation, and there is usually a decrease in the arterial-venous difference in partial pressure of oxygen.

Cyanide poisoning should be considered in any acyanotic patient who otherwise seems to suffer from severe hypoxia. The smell of bitter almonds on the breath or in gastric washings can be helpful. Plasma lactate levels are increased considerably. Whole blood cyanide levels often take hours to days for analysis due to unavailability of the analysis in most hospitals. For all but the mildest exposures, IV sodium thiosulfate provides sulphur substrate to facilitate the formation of thiocyanate by rhodanese. Inhaled amyl nitrite initiates the formation of methaemoglobin, ferric ion having an affinity for cyanide. The cyanide is then slowly released from the methaemoglobin and detoxified by the body. Intravenous sodium nitrate continues this process and may have additional beneficial cardiovascular effects. Extreme caution should be exercised in the formation of methaemoglobin in the patients also suffering from smoke inhalation.

Di-cobalt edetate and hydroxocobalamin are two antidotes common in Europe, based on the affinity of cyanide for cobalt. Supportive treatment measures include prompt removal from exposure, high concentration oxygen, correction of acidosis, treatment of seizures, and cardiovascular support. A high fatality rate is reported with cyanide reaching potentially 100%. However, cyanide being lighter than air, its volatility limits its toxicity in the open air. Most survivors are without sequelae, although anoxic encephalopathy has resulted in some long-term neurological effects [121].

6.4.6 Semicarbazide Hydrochloride

The food contaminant semicarbazide acts as an endocrine disrupter [122]. Semicarbazide (SEM) is a by-product of azodicarbonamide (ADC), which is used to foam the plastic gaskets of metal lids. In the food-production chain, SEM is released during manufacture when packaged foods are heated to ensure a tight seal, and it is present as food contaminant in baby foods, fruit juices, jams, and preserves. Moreover, SEM was also found in different foodstuffs such as powder and liquid milk, egg, and whey powder [123, 124].

SEM appeared to act as an endocrine disrupter showing multiple and gender specific mechanisms of action(s). The most vulnerable and susceptible phases of life cycle to ED adverse effects are pregnancy, childhood, and puberty when the endocrine system plays a key role in the development and differentiation of the whole organism. In humans, EDs have been mainly associated with apparent changes in both male and female reproductive health [125].

6.4.7 Food Additives

In the broadest of terms, food additives are substances intentionally added to food either directly or indirectly with one or more of the following purposes: (1) To maintain or improve nutritional quality; (2) To maintain product quality and freshness; (3) To aid in the processing or preparation of food; and (4) To make food more appealing. Some 2,800 substances are currently added to foods for one or more of these uses. During normal processing, packaging and storage, up to 10,000 other compounds can find their way into food.

The main categories of additives are colors (e.g. E100, curcumin), preservatives (e.g. E200, sorbic acid); antioxidants (e.g. E300, L-ascorbic acid); emulsifiers and stabilizers (e.g. E322, lecithins); and sweeteners (e.g. E421, mannitol). Other food additives include [126]:

- acids (e.g. citric acid, give a sour taste).
- anti-caking agents (e.g. some phosphates, to help food flow easily).
- antifoaming agents (e.g. oxystearin, to prevent excessive frothing).
- bases (e.g. bicarbonate, as a raising agent and acid neutralizer).
- bulking agents (e.g. guar gum, adds bulk without adding any calories).
- firming agents (e.g. aluminium salts, to retain crispness).
- flavour modifiers (reduces flavour).
- flour improvers (e.g. cysteine).
- glazing agents (e.g. waxes, to give polished appearance).
- humectants (e.g. glycerol, to prevent foods, such as marshmallow, drying out).
- liquid freezants (e.g. liquid nitrogen, to freeze food quickly).
- packaging gases (e.g. nitrogen, to control the atmosphere within a package).
- propellants (e.g. carbon dioxide, to form an aerosol, forcing food out of containers).
- release agents (e.g. silicates, to prevent food sticking to pans).
- sequestrants (e.g. sodium hydrogen diacetate, to help remove heavy metals from food).
- solvents (e.g. glycerol, to dissolve solids in food).

6.4.7.1 Health Hazards of Certain Food Additives

Some food additives are neurotoxic, which means they're capable of altering the normal activity of the nervous system – and even killing neurons. Symptoms include: Limb weakness or numbness; Loss of memory, vision, and intellect; Headache; Cognitive and behavioral problems; Sexual dysfunction. Major concerns from certain food additives include allergens, illness, birth and developmental defects, reproductive abnormalities, and cancer development [127].

A report published at the website (**URL: www.SixWise.com**), deals with **"12 Dangerous Food Additives: The dirty dozen food additives you really need to be aware of";** based on they have adverse effects to human health including cancer. These 12 substances are:

Propyl gallate; BHA (butylated hydroxyanisole) & BHT (butylated hydroxytoluene); Potassium bromate; Monosodium glutamate (MSG); Aspartame (Equal, NutraSweet); Acesulfame-K; Olestra; Sodium nitrite; Hydrogenated vegetable oil; Blue 1 & Blue 2; Red 3; and Yellow 6.

6.4.8 Melamine Contamination

Due to recent concerns of melamine-adulterated foods, the subject will be presented in more details.

Melamine (2,4,6-triamino-1,3,5-triazine) is an organic compound, a base with the chemical formula C3H6N6, was first prepared in 1834 by Liebig by heating potassium thiocyanate with ammonium chloride. The manufacture of melamine on a large scale is based on heating dicyandiamide or urea in the presence of ammonia [128]. Melamine can be hydrolyzed to ammeline, ammelide, and cyanuric acid. Besides, those derivatives are by-products formed during melamine synthesis. The overall reactions and the structures of all the compounds are shown in Fig. 6.1 [129].

Melamine is a common organic compound that is often combined with formaldehyde to be used in the manufacturing of plastics, including whiteboards, kitchenware, and commercial filters. Melamine resins are known as thermoset plastic, because the plastic is fixed after molding. If exposed to enough heat, melamine resins will melt. Melamine is also used in a wide range of flame-resistant materials. It protects wearers from heat hazards and helps to resist the spread of fire in aircraft and buses by acting as a fire blocker [129].

A major food safety incident in China was made public in September 2008. Kidney and urinary tract effects, including kidney stones, affected about 300,000 Chinese infants and young children, with six reported deaths. The source of the illness was traced to the contamination of infant formula with melamine. Investigations showed that melamine had been deliberately added to diluted raw milk to boost its apparent protein content [130]. Previous outbreaks of renal failure related to melamine have been reported in pets in 2004 in the Republic of Korea and in 2007 in the United States when the substance was added deliberately to a pet food ingredient [131]. Because melamine is cheap and easily accessible, there is an economic incentive for its (illegal) addition. Melamine's nitrogen level is high, 66% nitrogen by mass, which confers the analytical characteristics of protein molecules. The reason for this type of adulteration is that the test for the protein content of certain foods and food ingredients is assayed through a nonspecific procedure, the Kjeldahl reaction [132]. The procedure quantifies the presence of nitrogen, which is present in amino acids and therefore in proteins, but is also present in many other nonproteinaceous molecules such as melamine [133]. Thus, the addition of melamine can lead to an incorrectly high protein reading. The manufacturers

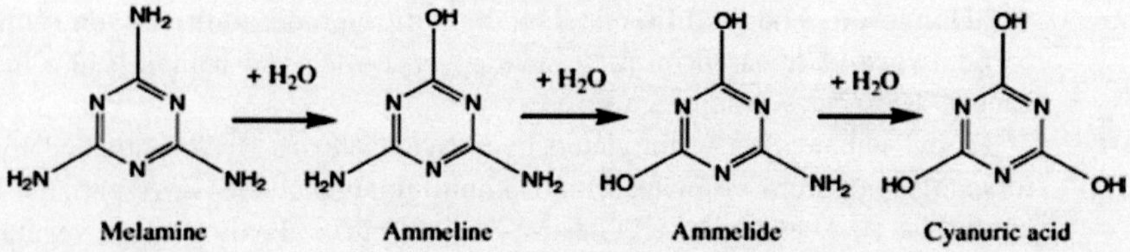

Fig. 6.1 Formation and structures of melamine and related compounds (Adopted from [129])

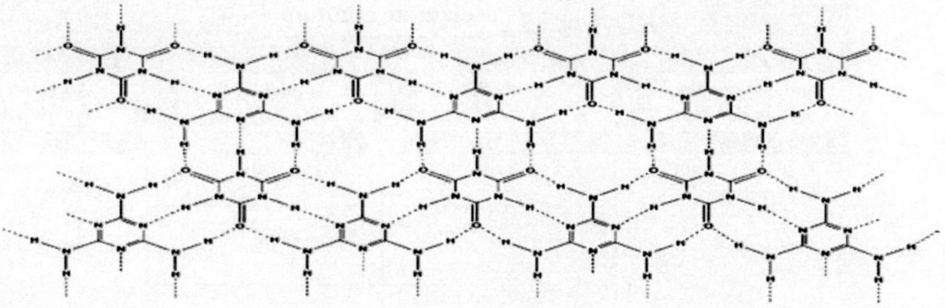

Fig. 6.2 The chemical structure of melamine-cyanoric acid crystal (Adopted from [129])

of the milk ingredient in China, in 2008, intentionally adulterated the product to give spuriously high readings for protein.

Thus, aside from common commercial uses, melamine became a topic of discussion in 2007 when veterinary scientists determined pet food contamination of melamine was the cause of hundreds of pet deaths. Prior to these reports, melamine had been regarded as nontoxic or minimally toxic. However, the unexplained presence of melamine in pet food is most likely the cause. The pet owners reported symptoms associated with renal failure, which could be explained by the ammonia that may result from the digestion of melamine. In addition, previous reports indicated the combination of melamine and cyanuric acid formed insoluble crystals in kidneys [134].

The LD50, the toxic dose, of melamine was more than 3 $g\,kg^{-1}$ body weight (**LD50–3,296** mg kg^{-1} for melamine, LD50>7,940 mg kg^{-1} for cyanuric acid) [135–137]. The contaminants are of concern, such as cyanuric acid, which might be found in the raw material. Melamine and melamine analogues (cyanuric acid, ammelide, and ammeline) are assumed to be of equal potency. It has been shown that a single oral exposure to a mixture of melamine and cyanuric acid can result in acute renal failure, which eventually caused the dog deaths. The chemical structure of the melamine-cyanuric acid crystal is shown in Fig. 6.2 [129].

Similar results were found in cats, dogs, and rats [134, 138, 139]. In 2007, the US Food and Drug Administration (FDA) identified and reported melamine and cyanuric acid as suspected contaminants in certain pet foods.

6.4.8.1 Sources and Levels of Contamination

The sources of melamine contamination have been divided into "baseline" levels, which refer to levels in food that do not result from adulteration or misuse, and "adulteration" levels, which refer to levels in food that result from the intentional illegal addition of melamine to food or feed [140]. Such a distinction is useful for practical purposes, but it is evident that a clear distinction is not always possible. For example, low levels of melamine in food could result from carryover from adulterated animal feed.

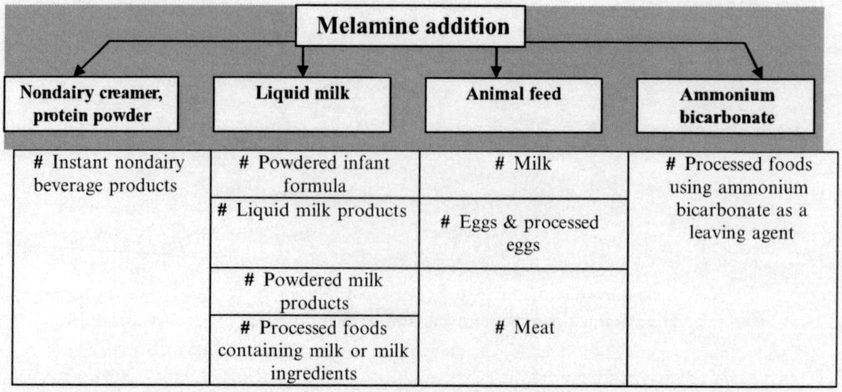

Fig. 6.3 Flow chart of the melamine-contamination chain from adulteration of milk and dairy products as observed during the 2008 incident in China (Adopted from [130])

Baseline concentrations of melamine are present in the environment and in the food chain as a result of the widespread use of materials that contain melamine. Generally, baseline levels are expected to be<1 mg/kg [140], and these levels are not considered to be a health concern. Melamine concentrations in food and animal feed above baseline levels are considered to be the result of misuse or adulteration. The high levels of melamine detected in many products related to the 2008 incident are a clear indication of adulteration.

Figure 6.3 presents the four major possible paths for deliberate contamination: (a) liquid milk in the milk-collecting stations that was then used in the production of powdered infant formula, liquid and powdered milk products, and processed milk-containing foods; (b) animal feed that resulted in contamination of milk, eggs, and potentially meat; (c) nondairy creamer and protein powder that lead to the contamination of instant nondairy beverage products; and (d) ammonium bicarbonate that was used to produce several types of processed food.

It is important to monitor all kinds of foods to prevent melamine contamination. Tyan et al. [129] reviewed the variety of analytical methods that have been used for the analysis of melamine in food. The limit of detection of these various methods is 0.05–100 ppm. The maximum acceptable concentration in food has been set at 50 ppb by the US FDA. A fast and ultrasensitive procedure for screening, detection, and characterization of melamine and its derivative compounds needs to be established. Currently, massspectrometry technologies provide an alternative to derivatization for regulatory analysis of food.

6.4.8.2 Toxicology and Risk Assessment of Melamine Contaminants

Several national and regional authorities around the world and the WHO have issued preliminary risk assessments and guidance on levels in food, mainly based on information from the 2007 pet-food incident, as a first pragmatic approach for public health protection [141]. Subsequently, a meeting of independent international scientific experts was organized by WHO and the following brief summary provides relevant aspects of the report [140].

Toxicology: Absorption, distribution, metabolism, and excretion. Melamine and its structural analog cyanuric acid are rapidly absorbed and rapidly excreted almost completely unmetabolized in the urine. The elimination half-lives for these two compounds are about 3 h [140, 142]. No information is available for other structural analogs, and no information is available on the absorption and excretion of the melamine-cyanurate complex.

Toxicity of melamine. Melamine is of low acute toxicity, with oral median lethal doses in mice and rats exceeding 3,000 mg/k body weight. Several subchronic studies in rats and mice are reported [143, 144], at doses up to 18,000 mg melamine/kg feed. The main observed toxicity was related to the excretory organs, kidney, and bladder. The most consistent and dose-related effects observed were bladder-stone formation and hyperplasia of the bladder epithelium. These effects were more pronounced in males than in females.

Melamine has been tested for carcinogenicity by oral administration in mice and rats [144]. The most significant and consistent finding from these studies was the production of urinary bladder carcinomas in male rats. In male mice, urinary bladder hyperplasia was detected. All tumor findings correlated strictly with calculus formation and exposure to high doses. In female rats, chronic inflammation of the kidney, distinct from chronic nephropathy in aging rats, was observed. Melamine is not genotoxic, and it is not considered a reproductive and developmental toxicant.

Toxicity of cyanuric acid. The toxicologic profile of cyanuric acid is very similar to that of melamine. It is of low acute oral toxicity, and subchronic toxicity at concentrations up to 5,375 mg/L in drinking water resulted in a low incidence of bladder calculi at the highest dose. In a 2-year study of rats that were given sodium cyanurate in the drinking water at doses estimated up to 371 mg/kg body weight per day (5,375 mg/L), no substance-related increase in tumor incidence was observed [142]. Cyanuric acid is not considered to be genotoxic, and it is not teratogenic or a reproductive toxicant.

Combined toxicity. From previous incidents in pets and livestock, and after experimental studies to investigate the combined effects of melamine plus cyanuric acid in cats [139] and in fish and pigs [145], it is apparent that oral exposure to melamine given simultaneously with cyanuric acid caused much more severe renal damage than did oral exposure to melamine or cyanuric acid alone.

Risk assessment. The formation of bladder calculi was identified as the most relevant end point, and because the calculi formation is dose-dependent or local-concentration dependant, with no signs of significant accumulation, the subchronic studies in rats serve as the basis for the risk assessment.

By applying dose–response modeling to the combined data for male rats from the two subchronic feeding studies, a 95% lower bound of the 10% benchmark dose (BMDL10) of 415 mg/kg diet was calculated. Dietary conversion and an additional feed-intake–reduction factor of 14% led to the BMDL10 of 35 mg/kg body weight per day. An uncertainty factor of 200 was applied, deriving a tolerable daily intake (TDI) of 0.2 mg/kg body weight (rounded to one significant figure). The uncertainty factor comprises the default 100-fold factor plus an extra 2-fold factor to account for sensitivity of infants and for data uncertainty in relation to possible underreporting of bladder calculi due to tissue preparation.

For adults, estimated exposure from "baseline" levels of melamine, defined as levels in food that do not result from adulteration or misuse from all sources, has been estimated to be up to 13 μg/kg body weight per day. Conservative exposure estimates from adulterated products from this incident were 0.8–3.5 times the TDI. For comparison, estimated exposure of infants in China to adulterated infant formula, at median levels of the most affected brand, ranged from 8.6 to 23.4 mg/kg body weight per day. These levels are about 40–120 times the TDI and explain the dramatic health outcome in Chinese infants.

In their Commentary, Gossner et al. [130], demonstrated the "International consequences and lessons learned" from melamine disaster. They mentioned that in any important food-safety event, there is an urgent need to provide the best available scientific knowledge in the area. Such knowledge can save lives and help to control an outbreak. Sharing information across borders is essential to obtain the best advice possible and to avoid confusion when tackling international events. This information should include international agreement about testing and reporting methodology; without these, coherent international analysis and action are not possible. In general, there is a need to achieve international scientific agreement relative to the risk of melamine in food and feed. A WHO expert meeting [140] provided the first international forum for exchange and joint analysis of data in this area. The Chinese authorities should be commended for the pertinent data provided at this meeting.

During the 2008 incident, significant confusion existed regarding which level of melamine in food presents a human health risk. Communication relative to the levels protective of health was made additionally difficult by the fact that countries in some cases published different "action" levels; some countries even used any melamine concentration detected as a signal for action. It is important for authorities to present clear, understandable reasoning for any action or nonaction taken. There is a need for a common understanding of the underlying science that ideally leads to one harmonized international set of limits in food and feed, typically achieved through the Codex Alimentarius Commission.

The melamine incident in September 2008 has clearly demonstrated a need to develop, at the international level, risk-based import-inspection systems. In addition, guidance regarding the necessary measures to demonstrate, with a certain degree of confidence, the appropriate levels of safety when a major food-safety event has seriously shaken confidence in the capabilities of a system to ensure the safety of the food it produces. This incident has also clearly demonstrated that food safety can only be ensured if all the stakeholders along the food chain are sharing information and data in a timely manner.

6.4.9 Chemical Substances of Natural Origin

These include bacterial secondary metabolites, fungal secondary metabolites (Mycotoxins), and *botanical substances.*

Toxins produced by microorganisms are chemical substances excreted in food contaminated with certain bacteria and fungi. They are extremely toxic and hazardous to human health [146]. Examples are given below:

- *Botulinum*: Botulinum neurotoxins (BoNTs) are produced by the Gram-positive anaerobic soil bacterium *Clostridium botulinum.* They are considered the most poisonous substances known to humans (LD50 = 1 ng/kg b.wt.), and are the cause of the life-threatening neuroparalytic disease botulism. BoNT-A is 100 billion times more toxic than cyanide [147].

  Due to the extreme potency, lethality and easy procurement, BoNTs have the potential to be very dangerous biological weapon and therefore represent significant warfare and terrorism threat [148]. Consequently, BoNTs are one of the six category A agents listed as the highest risk threat agents for bioterrorism by the US Centers for Disease Control and Prevention (CDC).

  Exposure to these toxins is generally through the consumption of contaminated canned vegetables, but almost every type of canned food has been associated with botulism.
- *Mycotoxins*: Mycotoxins are toxic substances produced mostly as secondary metabolites by fungi that grow on seeds and feed in the field, or in storage. Mycotoxins are important because they can be costly when they affect: animal productivity, human health, and international trade. Mould contamination usually occurs in the field, and mycotoxins can develop at various stages (e.g., preharvest growth, harvest, storage).

  Mycotoxins can cause different toxic effects depending on the type of toxin and the dosage. High doses of mycotoxin can cause acute illness or death; low doses cause chronic toxicity. The toxicity of mycotoxins may affect the body's nervous, heart, lungs, and digestive tract. Mycotoxins are also associated with acute hepatitis and liver cancer.

  The 2004 aflatoxin-poisoning outbreak in Kenya was the largest and most severe documented worldwide. The outbreak covered more than seven districts and resulted in 317 case-patients and 125 deaths. Other smaller outbreaks were reported in 2005 and 2006, with about 30 and 9 deaths, respectively. The maize implicated in the 2004 aflatoxicosis outbreak was harvested in February during off-season, early rains. Maize obtained from local farms in the affected area was significantly more likely to have aflatoxin levels greater than the regulatory limit of 20 ppb compared to maize bought from other regions of the country [149]. The following outlines health implications due to consumption of mycotoxin-food contaminants:

- The major mycotoxin producing fungi are species of Aspergillus, Fusarium and Penicillium and the important mycotoxins are aflatoxins, fumonisins, ochratoxins, cyclopiazonic acid, deoxynivalenol/nivalenol, patulin, and zearalenone. The food-borne mycotoxins likely to be of greatest significance for human health in tropical developing countries are aflatoxins and fumonisins.
- Mycotoxin contamination in certain agricultural systems has been a serious concern for human and animal health [150]. They have been reported to be carcinogenic, teratogenic, tremorogenic, haemorrhagic and dermatitic to a wide range of organisms, and known to cause hepatic carcinoma in man [151, 152]. In a normal varied human diet, constant exposure to low levels of several toxins is possible.
- Very little is known about the effects of long-term low-level exposure, especially with regard to co-contamination with multiple mycotoxins. Sedmikova et al. [153] found that ochratoxin A could increase the mutagenicity of aflatoxin B1 in the case of their simultaneous occurrence in the same substrate.
- Drinking or non-drinking water may be an effective medium for mycotoxin dispersal as a weapon. The threat from contaminated drinking water is obvious. In the case of non-drinking water, the toxin could be spread by spray from a shower and then breathed in. Work places where a lot of water may be employed, such as farms, or a car wash, could be susceptible. The levels of safety required for water for livestock may be considerably lower than water for human consumption, so this could be a potential route of attack [154].
- Mycotoxins have been ranked as the most important chronic risk factor in the diet above pesticide residues, synthetic contaminants, plant toxins, and food additives. They are considered more acutely toxic than pesticides. They are the biggest chronic health risk when incorporated into the diet [155].
- The current list of fungal toxins as biochemical weapons is small, although awareness is growing of the threats they may pose.
- **Trichothecenes** are a large family of chemically related mycotoxins produced by various species of Fusarium, Myrothecium, Trichoderma, Trichothecium, Cephalosporium, Verticimonosporium, and Stachybotrys. All mycotoxins in this group have small molecular weights, the same basic ring structure, and a characteristic 12,13-epoxide group [156].
- Trichothecene mycotoxins have an immunosuppressive effect on the health of animals and humans due to their multiple inhibitory effects on eukaryotic cells, including inhibition of protein, DNA and RNA synthesis, inhibition of mitochondrial function, and effects on cell division and membrane function [157].
- The main sources of trichothecenes in human food and animal feed are wheat, barley, and maize. Trichothecenes can enter human and animal food chains through breakfast cereals, bakery products, snack foods, beer, and compound feeds made from small grains and maize [158, 159]. In addition,

trichothecenes have the potential to enter human and animal food chains through milk, meat and eggs from livestock and poultry animals that are fed with trichothecene-contaminated feed, although the exposure risk to humans by consumption of edible tissues of animals exposed to trichothecenes is minor compared to the direct consumption of grain products [160–162].

- *Ricin*: Ricin is a protein cytotoxin derived from castor beans. Worldwide, one million tons of castor beans are processed annually in the production of castor oil; the waste mash is 5% ricin by weight [163]. Ricin can be prepared in liquid or crystalline form in large quantities without great technological capacity or expense. It is stable under ambient conditions and has high media notoriety. As an attack agent, it could be used to contaminate food or limited water supplies or aerosolised for inhalation. Clinical manifestations of ricin intoxication vary with the route of exposure. Ingestion of ricin by humans is expected to cause necrosis of the gastro-intestinal epithelium, local haemorrhage and hepatic, splenic and renal necrosis. Intramuscular (i.m.) injection is expected to cause severe local necrosis of muscle and regional lymph nodes with moderate visceral organ involvement. Accidental inhalational exposure, after 4–8 h, produced fever, chest tightness, cough, dyspnoea, nausea and arthralgias. It is anticipated that death could occur as a consequence of high permeability pulmonary edema, adult respiratory distress syndrome, and respiratory failure.

  Ricin intoxication is diagnosed by clinical suspicion in the correct epidemiological setting. Specific antigen testing on serum and respiratory secretions and immunohistochemical staining of tissues are useful where available; if not, seroconversion can be demonstrated. Treatment is supportive. The particulate filter mask is effective in preventing aerosol exposure. Candidate vaccines are being developed.
- *Abrin*: Abrin is a similar toxin to ricin, found in the highly ornamental "rosary pea." It previously used as a chemical/biological warfare agent during the First and the Second World Wars.

  Abrin induces oxidative stress and DNA damage in U937 human myeloleukemic cells [164].

6.5 Deliberate Contamination of Food

For the purposes of this document, food terrorism is defined as: an act or threat of deliberate contamination of food for human consumption with biological, chemical, and physical agents or radionuclear materials for the purpose of causing injury or death to civilian populations and/or disrupting social, economic, or political stability. Contamination of food and water supplies for terrorist purpose is real and current threat. The major concern of any government all over the world is to protect its inhabitance and provide them with healthy environment and safe food and water supplies by putting new roles which helps to prevent or detect any possible contamination through food handling during processing and other services [30].

The biological agents referred to are communicable infectious or non-infectious pathogenic microorganisms, including viruses, bacteria, and parasites. The chemical agents in question may be man-made or natural toxins. Physical agents can include a wide range of objects including glass, needles, and metal fragments. Radionuclear materials are defined in this context as radioactive chemicals capable of causing injury when present at unacceptable levels.

The objective of terrorists in using such agents against a civilian population is essentially the same as that of their use in warfare against military targets: to cause widespread incapacitation and injury and/or to effect terror and panic. Civilian populations are usually more vulnerable than military personnel to chemical, biological, or radionuclear weapons because they are of all ages and states of health, whereas military personnel are generally healthy adults. Furthermore, the latter are usually prepared for attack by training and in many cases protected by immunization, [prophylactics] and protective clothing and devices. The potential agents and circumstances of terrorist attacks in civilian settings are more diverse than those directed at military personnel. As a result, rapid diagnosis and appropriate, readily available treatment may be difficult to assure. Because of this diversity, the agents used by terrorists may be more easily obtainable than those used against military personnel [165].

6.5.1 The Agents May Terrorist Use in His Attack

Terrorist may use biological agents such as bacteria and viruses, or chemical agents which can be characterized as chemical warfare agents such as nerve, blood, and choking agent, also may use physical agents which can cause adverse health effects if eaten such as bone, glass fragments, metals. All these agents can be delivered as liquid, aerosols, or solid.

6.5.2 Chemical Agents

A chemical attack is the spreading of toxic chemicals with the intent to do harm to people or to environment on our planet. A wide variety of chemicals could be made, stolen, or otherwise acquired for use in an attack. Industrial chemical plants or the vehicles used to transport chemicals could also be sabotaged. Harmful chemicals that could be used in a terrorist attack include:

1. Chemical weapons (warfare agents) developed for military use.
2. Toxic industrial and commercial chemicals that are produced, transported, and stored in the making of petroleum, textiles, plastics, fertilizers, paper, foods, pesticides, household cleaners, and other products.
3. Chemical toxins of biological origin such as ricin.

The toxicity of chemicals varies greatly: some are acutely toxic (cause immediate symptoms); others are not very toxic at all.

6.5.3 *How Toxic Chemicals Can Be Used in a Terrorist Attack?*

The release of toxic chemicals in closed spaces could deliver doses high enough to injure or kill large number of people. In open areas, a toxic chemical cloud (plume) would become less concentrated as it spreads and would have to be released in huge amounts in order to produce large number of casualties. Potential delivery methods of toxic chemicals include:

1. Ventilation systems of a building.
2. Misting, aerosolizing devices, or sprayers.
3. Passive release (container of chemical left open).
4. Bombs, mines, or other explosive devices that contain chemicals other than those used to create the explosion.
5. Improvised chemical devices that combine readily available chemicals to produce a dangerous chemical.
6. Sabotage of plants or vehicles containing chemicals.
7. Introduction of toxins in the food and water supply.

6.5.4 *Exposure Through Contaminated Food*

Chemical agents can make food highly toxic, sometimes without changing the appearance or taste of the food. Butter, oils, fatty meats, and fish absorb nerve agents so readily that removal of the agents is virtually impossible. Foods in bottles cans, or wrappings are not affected by agent vapor and can be salvaged following decontamination. The food supply is vulnerable to intentional contamination by toxins such as botulinum toxin [165].

6.6 Preventive Measures throughout Sophisticated Food Chain

According to Sekheta et al. [165], the following demonstrates the possible terrorist threats or attacks for the main links in the conventional food chain starting from farms and ending with final consumers, and possible preventive measures.

6.6.1 *Farm Level*

Good agriculture practice (GAP) where general risk prevention is advocated. These can be considered as prerequisite programs for full adoption to HACCP systems or programs. HACCP can be used to implement good agriculture practices, which may reduce or eliminate food safety risk. Agriculture production areas can be vulnerable to deliberate contamination, such as with highly toxic pesticides and other chemicals. Irrigation water can be easily contaminated with chemical or biological agents. Subsequent processing may provide HACCP where contamination can be detected and controlled. Security measures, such as control animal feed ingredients and safety assurance systems should be included in the quality control of such ingredients or additives. Farms have to develop a plan for isolation, cleaning and disinfects ion. A record for new animals' feed and other products is a must. Security measures should be also considered during the rest of the processes such as manufacturing, transporting, and distributing.

6.6.2 Storage and Transportation

There are several precautions which can be taken; physical measures, such as fencing and locks which can be used to secure storage facilities and transport containers. Transportation includes transport food from farm to factories and the distribution chains to customers and restaurants.

6.6.3 Processing

Prevention in such area starts from improving onsite security programs, such as restricting rights of entry and exit and locking up storage bulks. Hazard Analysis and Critical Control Points (HACCP) system can be implemented to prevent deliberate contamination into processing operations. The introduction of raw materials into the processing stream is a Critical Control Point (CCP) in the most of processing operations. Water is one of the most important raw materials used in food manufacturing and in food processing (washing and cleaning purpose), thus, precautions have to be taken including water analysis, and it is very often a Critical Control Point CCP. Air systems are also very important and can be serious source of food contamination. Heat treatment is a step in food processing, and it is often a Critical Control Point for microbiological contamination. Security in processing area is a must and should be monitored and controlled. Written plans and procedures have to be developed, with crises management teams that can quickly assess the scope of potential problems and contain them efficiently and promptly [166].

6.6.4 Wholesale and Distribution

Wholesale establishments and markets are among the most vulnerable parts of the food supply system. Inspection of incoming ingredients, compressed gas, packaging, and labels are very important. Controlled access and increased vigilance including security are required. More secure containers for bulk foods and the use of prepackaged materials are useful to prevent deliberate contamination. Buyers should be suspicious of being sold food at much less prices or from outside the normal distribution systems. There is a very important and difficult area which related to the employees of food industry, background checks including their mental health [167] will be of a great help.

6.6.5 Food Services and Restaurants

Condiments in open containers used in restaurants and institutional are vulnerable to deliberate contamination. In all restaurants there should be a hazard identification and risk assessment. Drivers who distribute food have to be well-known, identified, and medically checked. Trucks have to be sealed or locked upon arrival. No un-authorized person can enter the restaurant through back doors. Storage area has to be monitored at all the time. The leftover food has to be destroyed immediately in a correct manner in accordance to a written procedure.

6.7 Recent Approaches in Food-Contaminant's Assay

There is a constant need to improve the performance of current diagnostic assays as well as develop innovative testing strategies to meet new testing challenges.

Recently, reliance has increased on computational toxicology methods for predicting toxicological effects when data are limited. Advances in molecular biology, identification of biomarkers, and availability of accurate and sensitive methods allow us to more precisely define the relationships between multiple chemical exposures and health effects, both qualitatively and quantitatively. Key research needs are best fulfilled through collaborative research. It is through such collaborations that resources are most effectively leveraged to further develop and apply toxicity assessment methods that advance public health practices in vulnerable communities [168].

The use of nanoparticles promises to help promote *in vitro* diagnostics to the next level of performance. Quantum dots (QDs), gold nanoparticles (AuNPs), and superparamagnetic nanoparticles are the most promising nanostructures for *in vitro* diagnostic applications. These nanoparticles can be conjugated to recognition moieties such as antibodies or oligonucleotides for detection of target biomolecules. Nanoparticles have been utilized in immunoassays, immunohistochemistry, DNA diagnostics, bioseparation of specific cell populations, and cellular imaging. Nanoparticles-based diagnostics may open new frontiers for detection of tumours, infectious diseases, bio-terrorism agents, and neurological diseases, to name a few. More work is necessary to fully optimize nanoparticles use for clinical diagnosis and to resolve some concerns regarding potential health and environmental risks related to their use. However, we envision further developments of nanoparticles-based diagnostics will yield unique assays with enhanced sensitivity and multiplexing capability for the modern clinical laboratory [169].

6.8 Challenges Facing Food Safety and Security

6.8.1 Climatic Changes

According to general consensus, the global climate is changing, which may also affect agricultural and livestock production. The expected variation regarding the climate includes changes in the intensity and in the distribution of rainfall, the elevation of the level of the oceans and a growing increase in the frequency and intensity of extreme climatic phenomena. The potential impact of climate change on food security is a widely debated and investigated issue. Amongst the issues identified are mycotoxins formed on plant products in the field or during storage; residues of pesticides in plant products affected by changes in pest pressure; trace elements and/or heavy metals in plant products depending on changes in their abundance and availability in soils; polycyclic aromatic hydrocarbons in foods following changes in long-range atmospheric transport and deposition into the environment; marine biotoxins in seafood following production of phycotoxins by harmful algal blooms; and the presence of pathogenic bacteria in foods following more frequent extreme weather conditions, such as flooding and heat waves.

The impact of this climate change on various aspects of human and animal health and welfare is a topic widely debated. However, the consequences of climate change for the food system, which comprises all the stages from "farm to fork" (mainly primary production, processing, transport and trading), have received less attention compared with other human and animal health and welfare issues [168].

6.8.2 Production of Biofuels from Agricultural Raw Materials

According to IPCC [170], the biofuel demand for the transport sector in 2030 is forecast to be 45–85 EJ, based on primary biomass, or 30–50 EJ based on fuels. The same source indicates a global potential regarding the energy supply of biomass of 125–760 EJ in the year of 2050. This makes the energy use of biomass, in its different variants, a subject to be considered as an important element towards the mitigation of the greenhouse effect.

Poverty in rural areas and the lack of programs and funding for agricultural developments are the most important causes of nourishment insecurity; conflicts, terrorism, corruption and environmental degradation also contribute significantly towards the problem [171]. Food production in the world has increased substantially. However, the insufficient household and national income, as well as natural or man-caused catastrophes have prevented the population from satisfying their basic nourishment needs. Considering the expected population growth, which by the year 2050 must reach about 9.200 billion inhabitants [172], the problems regarding hunger and nourishment insecurity must continue or even increase dramatically in some regions, unless urgent measures are taken.

Biofuels are renewable, and they come from agricultural products such as sugarcane, oleaginous plants, forest biomass, and other sources of organic matter. They can be used either isolated or added to conventional fuels in blends. As examples, it is possible to mention biodiesel, ethanol, methanol, methane, and charcoal.

Ethanol, for example, can be produced out of any organic matter of biological origin that has considerable amounts of sugars and materials that can be converted into sugar such as starch or cellulose. Sugarcane, sugar beetroot, sugar sorghum are examples of raw material that contain sugar and, therefore, can be used for ethanol production. Wheat, barley, corn are also raw materials that contain starch, which can easily be converted into sugar by using the available technologies. A significant part of the wood of trees and herbs is composed by cellulose, which can also be converted into sugar, but the process is more complicated than the one required for starch.

It is important to keep in mind that in spite of the advantages that characterize the use of biofuels, their production and end use may have serious environmental impacts such as the use of large amounts of water, the destruction of forests, the reduction in food production and the increase in soil degradation [173].

6.9 Summary Notes and Concluding Remarks

- *General*:
  - **According to WWF-UK** [61]:
    - more than 300 man-made chemicals are present in human bodies;
    - all these chemicals are present in the fetus;
    - man-made chemicals are more dangerous than naturally occurring ones;
    - presence of these chemicals is dangerous even at low levels;
    - pollutants are found in intensively farmed food; and
    - man-made chemicals are linked to birth defects in humans.
  - **According to WHO** [174]:
    - Chemicals find their way into food through intentional use or through contamination during production, processing and preparation.
    - Chemicals can pose health risks to the populations if their presence exceeds their safety thresholds.
    - Responsibility for monitoring the food supply for chemicals rests with governments.
    - While the sampling and analysis of food can be expensive, the most cost-effective method for monitoring the food supply is the total diet study.
    - Periodic total diet studies can provide general assurance that the food supply is safe from potentially toxic chemicals.
    - Total diet studies can be used as a priority setting tool to assess whether contaminants as well as nutrients are at safe levels in the diet.

- **Miscellaneous**
  - Safe foods is composed of various work packages that include (1) the use of advanced analytical methodologies to study potential effects of agricultural practices on crop composition; (2) emerging risks in food safety; (3) assessment of consumer exposure to food safety hazards by the use of advanced statistical techniques; (4) confidence of consumers and other stakeholders in risk management in food safety; and (5) institutional arrangements for food safety governance. The findings of all these work packages are integrated into a new model on risk analysis, which will be refined with inputs solicited from stakeholders [175].
  - By using Technology of Remote Sensing (TRS), it may be possible to develop "indicators" for emerging hazards and risks to food safety.
  - Climatic changes may adversely increase pest outbreaks and subsequent excessive use of pesticides. On the other side, climatic changes may decline production of sensitive economic crops; changing crop map over the earth planet.

- *Specific*:
  - As a matter of fact, the possibility of contaminating food and water supplies deliberately by a terrorist attack must be taken seriously. The key to preventing from terrorist attacks is coming from improving quality control and implementing a reasonable security measures at production facilities based on vulnerability assessment. There may not be an optimal system for all food businesses at all stages along the sophisticated food chain but current Hazard Analysis and Critical Control Point (HACCP) approaches have clear benefits [165].
  - Intentional contamination of the food supply by terrorists is a new threat with unique challenges. Thus, it requires increased food inspection, disease surveillance, laboratory capacity, and awareness among health professionals and the general public.
  - Globalization of food trade is considered to be able to contribute to more widespread dissemination of hazards. From another perspective, however, globalization may increase the sources of food available to affluent urban populations, which increases their food entitlements and thereby decreases their vulnerability to shock in the food supply [175].
  - Reducing the risk of terrorists will require an unprecedented degree of cooperation among the public health, government, utilities, commercial, and other private sectors. The WHO guidelines [141] assist Member States on preventing terrorist.
  - All countries must have basic systems to prevent deliberate contamination of their food supplies and, if attacked, to respond rapidly to minimize the health, economic, and other effects. Cooperation between countries has to be activated, and exchanging information is something very important to minimize threat and contamination on national, regional, and international levels.
  - Prevent and respond to the food safety emergencies must be close to effective responses. It is everyone's responsibility all over the world in such manner. Prevention and response are the two major strategies for countering the threat of food terrorist [141].

References

1. FDA (2008) Animal cloning: a risk assessment. U.S. Food and Drug Administration. Center for Veterinary Medicine. Department of Health & Human Services. http://www.fda.gov/bbs/topics/NEWS/2006/NEW01541.html
2. Fries GF (1995) A review of the significance of animal food products as potential pathways of human exposures to dioxins. J Anim Sci 73(6):1639–1650
3. Tricker AR, Preussmann R (1990) Chemical food contaminants in the initiation of cancer. Proc Nutr Soc 49:133–144
4. Croci L, De Santis B, Dekkers S, Filippi L, Hutjes RWA, Noordam MY, D'Mello JPF (2003) Food safety: contaminants and toxins. CABI, Wallingford/Cambridge
5. Spears M, Gregorie M (2004) Foodservice organizations: a managerial and systems approach, 5th edn. Pearson Education, Upper Saddle River
6. Payne-Palacio J, Theis M (2001) West & Wood's introduction to foodservice, 9th ed. Prentice-Hall, Upper Saddle River. http://www.foodtimeline.org/restaurants.html. ©Lynne Olver 2000. Accessed 6 June 2010
7. Reynolds D (2003) On-site foodservice management. Wiley, Hoboken
8. FDA (2003) Guidance for industry: retail food stores and foodservice establishment: food security preventive measures. U. S. Food and Drug Administration (November 2003). Retrieved from http://www.cfsan.fda.gov/guidance.html
9. Biberoglu S, Biberoglu K, Komsuoglu G (1988) Mad honey. J Am Med Assoc 259:1943
10. Head T, Landes R (1994) The peace of God: social violence and religious response in France around the year 1000. Speculum 69:163–169
11. Woolf A (2000) Witchcraft or mycotoxin? The Salem witch trials. J Toxicol Clin Toxicol 38:457–460
12. Emmerson BT (1970) "Ouch-ouch" disease: the osteomalacia of cadmium nephropathy. Ann Intern Med 73(5):854–855
13. Bakir F, Damluji SF, Amin-Zaki L, Murtadha M, Khalidi A, Al-Rawi NY, Tikriti S, Dhahir HI, Clarkson TW, Smith JC, Doherty RA (1973) Methyl mercury poisoning in Iraq. Science 181:230–241
14. Smith H, Spalding DM (1959) Outbreak of paralysis in Morocco due to orto-cresyl phosphate poisoning. Lancet 274(7110):1019–1021
15. Barceloux DG (2008) Medical toxicology of natural substances: foods, fungi, medicinal herbs, toxic plants, and venomous animals. Wiley, Hoboken, 1158 pp
16. Greenpeace and WWF-UK (2005) A present for life: hazardous chemicals in umbilical cord blood, Greenpeace Netherlands/Greenpeace International/WWF-UK, Sept 2005, ISBN 90-73361-87-7, 59 pp. http://www.wwf.org.uk/filelibrary/pdf/presentforlife.pdf
17. Colborn T, Myers J, Dumanoski D (1996) Our stolen future. Dutton, NewYork, ISBN 0-525-93982-2
18. Srivastava S, Awasthi VK, Srivastava SP, Seth PK (1989) Biochemical alterations in rat fetal liver following in utero exposure to di (2-ethylhexyl) phthalate (DEHP). Indian J Exp Biol 27(10):885–888
19. Skakkebaek NE, Meyts R-D, Main KM (2001) Testicular dysgenesis syndrome: an increasingly common developmental disorder with environmental aspects. Hum Reprod 16(5):972–978
20. Harrad S, Wijesekera R, Hunter S, Halliwell C, Baker R (2004) Preliminary assessment of U.K. human dietary and inhalation exposure to polybrominated diphenyl ethers. Environ Sci Technol 38(8):2345–2350
21. Greenpeace-UK (2003) Consuming chemicals – hazardous chemicals in house dust as an indicator of chemical exposure in the home. www.greenpeace.to/publications/housedust_uk_2003.pdf
22. Greenpeace (2005) Perfume – an investigation of cChemicals in 36 Eaux de Toilette and Eaux de Parfum, Greenpeace International, Feb 2005, 16 pp. http://www.greenpeace.org/raw/content/international/press/reports/perfume-an-investigation-of.pdf

23. Okeeffe M, Kennedy O (1998) Residues – a food safety problem? J Food Saf 18:297–319
24. Rhodehamel EJ (1992) Overview of biological, chemical, and Physical hazards. In: Pierson MD, Corlett DA Jr (eds) HACCP principles and applications. Chapman & Hall, New York, pp 8–29
25. Ropkins K, Beck AJ (2003) Using HACCP to control organic chemical hazards in food wholesale, distribution, storage and retail. Trends Food Sci Technol 14:374–389
26. Addis PB, Hassel CA (1992) Safety issues with antioxidants in foods. In: Finley JW, Robinson SF, Armstrong DJ (eds) Food safety assessment, vol 484, ACS symposium series. American Chemical Society, Washington, DC, pp 346–376
27. Fleming HP, Daeschel MA, McFeeters RF, Pierson MD (1989) Butric-acid spoilage of fermented cucumbers. J Food Sci 54:636–639
28. Bánáti D (2008) Fear of food in Europe? Fear of foods in Europe through Hungarian experience. Trends Food Sci Technol 19:441–444
29. Eurobarometer (2006) Risk issues. Special Eurobarometer 238/Wave 64.1 -TNS Opinion & Social: 57pp + Annexes
30. Khan AS, Swerdlow DL, Juranek DD (2001) Precautions against biological and chemical terrorism directed at food and water supplies. Public Health Rep 116(1):3–14
31. Trewavas A (2004) A critical assessment of organic farming and food assertions with particular respect to the UK and the potential environmental benefits of no-till agriculture. Crop Prot 23:757–781
32. Gonzalez M, Miglioranza KSB, Aizpún de Moreno JE, Moreno VJ (2003) Occurrence and distribution of organochlorine pesticides (OCPs) in tomato (Lycopersicon esculentum) crops from organic production. J Agric Food Chem 51(5):1353–1359
33. Gonzalez M, Miglioranza KSB, Aizpún de Moreno JE, Moreno VJ (2005) Evaluation of conventionally and organically produced vegetables for high lipophilic organochlorine pesticide (OCP) residues. Food Chem Toxicol 43:261–269
34. Salim A (2006) Evaluation of heavy metal contents and organochlorine pesticides (OCPs) residues in Egyptian organically-farmed vegetables. J Agric Sci Mansoura Univ 31(3):1601–1612
35. Zohair A, Salim A, Soyibo AA, Beck AJ (2006) Residues of polycyclic aromatic hydrocarbons (PAHs), polychlorinated biphenyls (PCBs) and organochlorine pesticides in organically-farmed vegetables. Chemosphere 63:541–553
36. Mansour SA, Belal MH, Abou-Arab AAK, Ashour HM, Gad MF (2009) Evaluation of some pollutant levels in conventionally and organically farmed potato tubers and their risks to human health. Food Chem Toxicol 47:615–624
37. Mansour SA, Belal MH, Abou-Arab AAK, Gad MF (2009) Monitoring of pesticides and heavy metals in cucumber fruits produced from different farming systems. Chemosphere 75:601–609
38. Schecter A, Papke O, Tung KC, Staskal D, Birnbaum L (2004) Polybrominated diphenyl ethers contamination of United States food. Environ Sci Technol 8(20):5306–5311
39. WWF-UK (2006) Chain of contamination: the food link. World Wide Fund. 10 pp. Website: wwf.org.uk
40. WWF-UK (2005) Generations X. http://detox.panda.org/news_publications/publications_detail.cfm?uxNewsID=25616
41. WWF-UK and Greenpeace (2005) A present for life: hazardous chemicals in umbilical cord blood. Text: Jacqueline Schuiling and Wytze van der Naald; 55 pp. www.greenpeace.org/chemicals
42. Friends of the Earth Europe (2005) Toxic inheritance – More than 300 pollutants in breast milk – time for a new chemicals policy. http://www.foeeurope.org/publications/2005/toxic_inheritance.pdf
43. Gearhart J, Posselt H (2006) Toxic at any speed. Chemicals in cars and the need for safe alternatives. The Ecology Center Report, 32 pp. (www.ecocenter.org/dust/ToxicAtAnySpeed.pdf)
44. WWF International Arctic Programme and WWF-DetoX (2006) Killing them softly: health effects in Arctic wildlife linked to chemical exposures. (http://assets.panda.org/downloads/arctic_report___8_pager.pdf)

45. WWF International Arctic Programme (2005) The tip of the iceberg: Chemical contamination in the Arctic. (http://assets.panda.org/downloads/the_tip_of_the_iceberg___summary.pdf)
46. European Commission (2006) Environment fact sheet: REACH a chemicals policy for the EU. http://ec.europa.eu/environment/chemicals/reach/fact_sheet.pdf
47. Mansour SA (2009) Persistent organic pollutants (POPs) in Africa: Egyptian Scenario. Hum Exp Toxicol 28(9):531–566
48. Anonymous (2001) Persistent organic pollutants and the Stockholm Convention: a resource guide. A report prepared by Resource Futures International for the World Bank and CIDA, 2001, 22 pp
49. WWF-UK (2004) Contamination: the next generation. http://www.wwf.org.uk/chemicals/publications.asp
50. Guenther K, Heinke V, Thiele B, Kleist E, Prast H, Raecker T (2002) Endocrine disrupting nonylphenols are ubiquitous in food. Environ Sci Technol 36(8):1676–1680
51. Harner T, Pozo K, Gouin T, Macdonald A-M, Hung H, Cainey J, Peters A (2006) Global pilot study for persistent organic pollutants (POPs) using PUF disk passive air samplers. Environ Pollut 144:445–452
52. Lichtenstein EP (1959) Absorption of some chlorinated hydrocarbon insecticides from soils into various crops. J Agric Food Chem 14:62–65
53. Mattina MJI, Ianucci-Berger W, Dykas L (2000) Chlordane uptake and its translocation in food crops. J Agric Food Chem 48:1909–1915
54. DHHS (1998) 8th Report on carcinogens, summary. US Department of Health and Human Services, Atlanta
55. Covaci A, Jorens P, Jacquemyn Y, Schepens R (2002) Distribution of PCBs and organochlorine pesticides in umbilical cord and maternal serum. Sci Total Environ 45–53:298
56. WWF (Apr 2004) Chemical check up – an analysis of chemicals in the blood of Members of the European Parliament, 48 pp. http://www.panda.org/detox
57. Dorner G, Plagemann A (2002) DDT in human milk and mental capacities in children at school age:an additional view on PISA 2000. Neuro Endocrinol Lett 23(5–6):427–431
58. Eriksson P (1997) Developmental neurotoxicity of environmental agents in the neonate. Neurotoxicology 18(3):719–726
59. Eskenazi B, Bradman A, Castorina R (1999) Exposures of children to organophosphate pesticides and their potential adverse health effects. Environ Health Perspect 107(suppl 3):409–419
60. WWF-UK (June 2004) Compromising our children – chemical impacts on children's intelligence and behaviour, 29 pp. www.wwf.org.uk/chemicals
61. WWF-UK (2003) Contamination: the results of WWF's Biomonitoring Survey. 32 pp. www.wwf.org.uk
62. Ohta S, Ishizuka D, Nishimura H, Nakao T, Aozasa O, Shimidzu Y, Ochiai F, Kida T, Nishi M, Miyata H (2002) Comparison of polybrominated diphenyl ethers in fish, vegetables, and meats and levels in human milk of nursing women in Japan. Chemosphere 46(5):689–696
63. Guvenius DM, Aronsson A, Ekman-Ordeberg G, Bergman A, Noren K (2003) Human prenatal and postnatal exposure to polybrominated diphenyl ethers, polychlorinated biphenyls, polychlorobiphenylols, and pentachlorophenol. Environ Health Perspect 111(9): 1235–1241
64. Hardell L, Lindstrom G, van Bavel B, Wingfors H, Sundelin E, Liljegren G (1998) Concentrations of the flame retardant 2, 2', 4, 4'- tetrabrominated diphenyl ether in human adipose tissue in Swedish persons and the risk for non-Hodgkin's lymphoma. Oncol Res 10(8):429–432
65. Schroter-Kermani (2001) Endocrine disrupters in human and environmental samples from the German Environmental Specimen Bank. In: Proceedings of the second status seminar: endocrine disrupters, Berlin, 2–4 Apr, 2001
66. Meironyte D, Noren K, Bergmann A (1999) Analysis of polybrominated diphenyl ethers in Swedish human milk. A time-related trend study, 1972–1997. J Toxicol Environ Health A 58:329–341

67. Hooper K, She J (2003) Lessons from the polybrominated diphenyl ethers (PBDEs): precautionary principle, primary prevention, and the value of community-based body-burden monitoring using breast milk. Environ Health Perspect 111(1):109–114
68. Mazdai A, Dodder NG, Abernathy MP, Hites RA, Bigsby RM (2003) Polybrominated diphenyl ethers in maternal and fetal blood samples. Environ Health Perspect 111(9):1249–1252
69. Thomsen C, Lundanes E, Becher G (2002) Brominated flame retardants in archived serum samples from Norway: a study on temporal trends and the role of age. Environ Sci Technol 36(7):1414–1418
70. Meerts IA, van Zanden JJ, Luijks EA, van Leeuwen-Bol I, Marsh G, Jakobsson E, Bergman A, Brouwer A (2000) Potent competitive interactions of some brominated flame retardants and related compounds with human transthyretin in vitro. Toxicol Sci 56(1):95–104
71. Olsen CM, Meussen-Elholm ET, Samuelsen M, Holme JA, Hongslo JK (2003) Effects of the environmental oestrogens bisphenol A, tetrachlorobisphenol A, tetrabromobisphenol A, 4-hydroxybiphenyl and 4, 4'-dihydroxybiphenyl on oestrogen receptor binding, cell proliferation and regulation of oestrogen sensitive proteins in the human breast cancer cell line MCF-7. Pharmacol Toxicol 92(4):180–188
72. Darnerud PO (2003) Toxic effects of brominated flame retardants in man and in wildlife. Environ Int 29(6):841–853
73. Darnerud PO, Eriksen GS, Johannesson T, Larsen PB, Viluksela M (2001) Polybrominated diphenyl ethers: occurrence, dietary exposure, and toxicology. Environ Health Perspect 109(suppl 1):49–68
74. Main KM, Mortensen GK, Kaleva MM, Boisen KA, Damgaard IN, Chellakooty M, Schmidt IM, Suomi A-M, Virtanen HE, Petersen JH, Andersson A-M, Toppari J, Skakkebaek NE (2006) Human breast milk contamination with phthalates and alterations of endogenous reproductive hormones in infants three months of age. Environ Health Perspect 114(2):270–276
75. Swan SH, Main KM, Liu F, Stewart SL, Kruse RL, Calafat AM, Mao CS, Redmon JB, Ternard CL, Sullivan S, Teague JL, Study for future families research team (2005) Decrease in anogenital distance among male infants with prenatal phthalate exposure. Environ Health Perspect 113(8):1056–1061
76. Dostal LA, Weaver RP, Schwetz BA (1987) Transfer of di (2-ethylhexyl) phthalate through rat milk and effects on milk composition and the mammary gland. Toxicol Appl Pharmacol 91(3):315–325
77. Parmar D, Srivastava SP, Srivastava SP, Seth PK (1985) Hepatic mixed function oxidases and cytochrome P-450 contents in rat pups exposed to di-(2- ethylhexyl)phthalate through mother's milk. Drug Metab Dispos 13(3):368–370
78. Rimkus G, Rimkus B, Wolf M (1994) Nitro musks in human adipose tissue and breast milk. Chemosphere 28(2):421–433
79. Barr DB, Silva MJ, Kato K, Reidy JA, Malek NA, Hurtz D, Sadowski M, Needham LL, Calafat AM (2003) Assessing human exposure to phthalates using monoesters and their oxidized metabolites as biomarkers. Environ Health Perspect 111(9):1148–1151
80. CDC (2003) Second national report on human exposure to environmental chemicals. (Revised version). Centres for disease control and prevention, National Center for Environmental Health, Atlanta. NCEH Pub. No. 02-0716. http://www.cdc.gov/exposurereport/pdf/secondNER.pdf. Accessed Sept 2003
81. Koch HM, Drexler H, Angerer J (2003) An estimation of the daily intake of di(2-ethylhexyl) phthalate (DEHP) and other phthalates in the general population. Int J Hyg Environ Health 206(2):77–83
82. Takada H, Isobe T, Nakada N, Nishiyama H, Iguchi T, Irie H, Mori C (1999) Bisphenol A and nonylphenols in human umbilical cords. In: Proceedings of the international scientific conference on environmental endocrine disrupting chemicals, Monte Verita, Ascona, 7–12 Mar 1999
83. Kafferlein HU, Angerer J (2001) Trends in the musk xylene concentrations in plasma samples from the general population from 1992/1993 to 1998 and the relevance of dermal uptake. Int Arch Occup Environ Health 74(7):470–476

84. Rimkus GG, Wolf M (1996) Polycyclic musk fragrances in human adipose tissue and human milk. Chemosphere 33(10):2033–2043
85. Bitsch N, Dudas C, Korner W, Failing K, Biselli S, Rimkus G, Brunn H (2002) Estrogenic activity of musk fragrances detected by the E-screen assay using human mcf-7 cells. Arch Environ Contam Toxicol 43(3):257–264
86. Schreurs RH, Quaedackers ME, Seinen W, van der Burg B (2002) Transcriptional activation of estrogen receptor ERalpha and ERbeta by polycyclic musks is cell type dependent. Toxicol Appl Pharmacol 183(1):1–9
87. Doering DD, Steckelbroeck S, Doering T, Klingmuller D (2002) Effects of butyltins on human 5alpha-reductase type 1 and type 2 activity. Steroids 67(10):859–887
88. Lo S, Allera A, Albers P, Heimbrecht J, Jantzen E, Klingmuller D, Steckelbroeck S (2003) Dithioerythritol (DTE) prevents inhibitory effects of triphenyltin (TPT) on the key enzymes of the human sex steroid hormone metabolism. J Steroid Biochem Mol Biol 84(5):569–576
89. Steckelbroeck S, Heidrich D, Heimbrecht J, Klingmuller D (2001) Effects of triphenyltin (TPT) on the key enzymes of the human sex steroid hormone metabolism. In: Proceedingof second status seminar: endocrine disrupters, Berlin, 2–4 Apr 2001, Germany, Abstracts, p127
90. Adeeko A, Li D, Forsyth DS, Casey V, Cooke GM, Barthelemy J, Cyr DG, Trasler JM, Robaire B, Hales BF (2003) Effects of in utero tributyltin chloride exposure in the rat on pregnancy outcome. Toxicol Sci 74(2):407–415
91. van Heijst APN (1994)Tributyltin compounds. International programme on chemical safety group poisons information monograph: G018. http://www.inchem.org/documents/pims/chemical/pimg018.ht. Accessed Sept 2003
92. Takahashi S, Mukai H, Tanabe S, Sakayama K, Miyazaki T, Masuno H (1999) Butyltin residues in livers of humans and wild terrestrial mammals and in plastic products. Environ Pollut 106(2):213–218
93. IPCS (1999)Tributyltin oxide. International programme on chemical safety concise international chemical assessment document: 14. http://www.inchem.org/documents/cicads/cicads/cicad14.htm#PartNumber:14. Accessed Sept 2003
94. Zaidi MI, Asrar A, Mansoor A, Farooqui MA (2005) The heavy metal concentrations along roadside trees of Quetta and its effects on public health. J Appl Sci 5(4):708–711
95. Trichopoulos D (1997) Epidemiology of Cancer. In: DeVita VT (ed) Cancer: principles and practice of Oncology. Lippincott, Philadelphia
96. Turkdogan MK, Kilicel F, Kara K, Tuncer I (2002) Heavy metals in soil, vegetables and fruits in the endemic upper gastrointestinal cancer region of Turkey. Environ Toxicol Pharmacol 13:175–179
97. Sharma RK, Agrawal M, Marshall FM (2008) Heavy metal (Cu, Zn, Cd and Pb) contamination of vegetables in urban India: a case study in Varanasi. Environ Pollut 154:254–263
98. Rangan C, Barceloux DG (2008) Medical toxicology of natural substances: foods, fungi, medicinal herbs, toxic plants, and venomous animals. In: Food contamination: chemical contamination and additives. Wiley, Hoboken, pp 5–21
99. Peterson J, MacDonell M, Haroun L, Monette F (2007) Radiological and chemical fact sheets to support health risk analyses for contaminated areas. Argonne National Laboratory, Environmental Science Division, Argonne (Aug). http://www.ead.anl.gov/pub/dsp_detail.cfm?PubID=1472
100. Goodson A, Summerfield W, Cooper I (2002) Survey of bisphenol A and bisphenol F in canned food. Food Addit Contam 19:1–12
101. Greenpeace Netherlands (November 2004) Chemical footprints in blood – the evidence. Produced by Stichting Greenpeace Nederland. Text: Jacqueline Schuiling: 28 pp
102. Schonfelder G, Wittfoht W, Hopp H, Talsness CE, Paul M, Chahoud I (2002) Parent bisphenol A accumulation in the human maternal–fetal–placental unit. Environ Health Perspect 110(11):A703–A707
103. Kuroda N, Kinoshita Y, Sun Y, Wada M, Kishikawa N, Nakashima K, Makino T, Nakazawa H (2003) Measurement of bisphenol A levels in human blood serum and ascetic fluid by HPL causing a fluorescent labeling reagent. J Pharm Biomed Anal 30(6):1743–1749

104. Stoker C, Rey F, Rodriguez H, Ramos JG, Sirosky P, Larriera A, Luque Munoz-de-Toro M (2003) Sex reversal effects on Caiman latirostris exposed to environmentally relevant doses of the xenoestrogen bisphenol-A. Gen Comp Endocrinol 133(3):287–296, Oct 1
105. Palanza P, Howdeshell KL, Parmigiani S, vom Saal FS (2002) Exposure to a low dose of bisphenol A during fetal life or in adulthood alters maternal behaviour in mice. Environ Health Perspect 110(suppl 3):415–422
106. Dessi-Fulgheri F, Porrini S, Farrabollini F (2002) Effects of perinatal exposure to bisphenol A on play behaviour of female and juvenile rats. Environ Health Perspect 110(suppl 3) :403–407
107. Kawai K, Nozaki T, Nishikata H, Aou S, Takii M, Kubo C (2003) Aggressive behavior and serum testosterone concentration during the maturation process of Male Mice: the effects of fetal exposure to bisphenol A. Environ Health Perspect 111(2):175–178
108. Kubo K, Arai O, Omura M, Omura M, Watanabe R, Ogata R, Aou S (2003) Low dose effects on bisphenol A on sexual differentiation of the brain and behaviour in rats. Neurosci Res 45(3):345–356
109. Rice D (2000) Parallels between attention deficit hyperactivity disorder and behavioral deficits produced by neurotoxic exposure in monkeys. Environ Health Perspect 108(suppl 3): 405–408
110. Ikezuki Y, Tsutsumi O, Takai Y, Kamei Y, Taketani Y (2002) Determination of bisphenol A concentrations in human biological fluids reveals significant early prenatal exposure. Hum Reprod 17(11):2839–2841
111. Howdeshell KL, Hotchkiss AK, Thayer KA, Vandenbergh JG, vom Saal FS (1999) Exposure to bisphenol A advances puberty. Nature 401(6755):763–764
112. Houlihan J, Wiles R, Thayer K, Gray S (2003) Body burden. The pollution in people. Environmental Working Group, Washington, DC. http://www.ewg.org/reports/bodyburden/. Accessed Sept 2003
113. Lopez-Cervantes J, Paseiro-Losada P (2003) Determination of bisphenol A in, and its migration from PVC stretch film used for food packaging. Food Addit Contam 20(6):596–606
114. Ozaki A, Baba T (2003) Alkylphenol and bisphenol A levels in rubber products. Food Addit Contam 20(1):92–98
115. Brede C, Fjeldal P, Skjevrak I, Herikstad H (2003) Increased migration levels of bisphenol A from polycarbonate baby bottles after dishwashing, boiling and brushing. Food Addit Contam 20(7):684–689
116. Inoue K, Kato K, Yoshimura Y, Makino T, Nakazawa H (2000) Determination of bisphenol A in human serum by highperformance liquid chromatography with multi-electrode electrochemical detection. J Chromatogr Biomed Sci Appl 749(1):17–23
117. Yamasaki H, Nagake Y, Makino H (2001) Determination of bisphenol A in effluents of hemodialyzers. Nephron 88(4):376–378
118. Uchida K, Suzuki A, Kobayashi Y, Buchanan DL, Sato T, Watanabe H, Katsu Y, Suzuki J, Asaoka K, Mori C, Arizono K, Iguchi T (2002) Bisphenol-A administration during pregnancy results in female exposure in mice and monkeys. J Health Sci 48:579–582
119. Takeuchi T, Tsutsumi O (2002) Serum bisphenol A concentrations showed gender differences, possibly linked to androgen levels. Biochem Biophys Res Commun 291(1):76–78
120. Eckstein M (2008) Enhancing public health preparedness for a terrorist attack involving cyanide. J Emerg Med 35(1):59–65
121. Bismuth C, Borron SW, Baudd FJ, Barriot P (2004) Chemical weapons: documented use and compounds on the horizon. Toxicol Lett 149:11–18
122. Maranghi F, Tassinari R, Marcoccia D, Altieri I, Catone T, De Angelis G, Testai E, Mastrangelo S, Maria Grazia Evandri MG, Bolle P, Lorenzetti S (2010) The food contaminant semicarbazide acts as an endocrine disrupter: Evidence from an integrated in vivo/ in vitro approach. Chem Biol Interact 183:40–48
123. Bendall JG (2008) Semicarbazide is non-specific as a marker metabolite to reveal nitrofurazone abuse as it can form under Hofmann conditions. Food Addit Contam 12:1–10

124. Maranghi F, Tassinari R, Lagatta V, Moracci G, Eusepi MA, Di Virgilio A, Scattoni ML, Calamandrei G (2009) Effects of the food contaminant semicarbazide following oral administration in juvenile Sprague–Dawley rats. Food Chem Toxicol 47:472–479
125. Mantovani A (2002) Hazard identification and risk assessment of endocrine disrupting chemicals with regard to developmental effects – Rev. Toxicology 181-182:367–370
126. Smith J, Hong-Shum L (eds) (2003) Food additives data book. Blackwell Science, Oxford
127. Maga JA, Tu AT (eds) (1994) Food additive toxicology. Marcel Dekker, New York
128. Bann B, Miller SA (1958) Melamine and derivatives of melamine. Chem Rev 58:131–172
129. Tyan Y-C, Yang M-H, Jong S-B, Wang C-K, Shiea J (2009) Melamine contamination. Anal Bioanal Chem 395:729–735
130. Gossner CM-E, Schlundt J, Embarek PB, Hird S, Lo-Fo-Wong D, Beltran JJO, Teoh KN, Tritscher A (2009) The melamine incident: implications for international food and feed safety. Environ Health Perspect 117(12):1803–1808
131. Brown CA, Jeong K-S, Poppenga RH, Puschner B, Miller DM, Ellis AE, Kang K-Il, Sum S, Cistola AM, Brown SA (2007) Outbreaks of renal failure associated with melamine and cyanuric acid in dogs and cats in 2004 and 2007. J Vet Diagn Invest 19:525–531
132. Kennaway EL (1921) The estimation of non-protein nitrogen in blood by a micro-Kjeldahl method. Biochem J 15:510–512
133. Abernethy DR, Sheehan C, Griffiths JC, Williams RL (2008) Adulteration of drugs and foods: compendial approaches to lowering risk. Clin Pharmacol Ther 85:444–447
134. Dobson RL, Motlagh S, Quijano M, Cambron RT, Baker TR, Pullen AM, Regg BT, Bigalow-Kern AS, Vennard T, Fix A, Reimschuessel R, Overmann G, Shan Y, Daston GP (2008) Identification and characterization of toxicity of contaminants in pet food leading to an outbreak of renal toxicity in cats and dogs. Toxicol Sci 106:251–262
135. Hammond BG, Barbee SJ, Inoue T, Ishida N, Levinskas GJ, Stevens MW, Wheeler AG, Cascieri T (1986) A review of toxicology studies on cyanurate and its chlorinated derivatives. Environ Health Perspect 69:287–292
136. IUCLID (2008) IUCLID chemical data sheet: substance ID 108-78-1, pp 30–98
137. OECD (Organisation for economic co-operation and development) (2002) SIDS Analysis. UNEP Publications, Melamine
138. Cianciolo RE, Bischoff K, Ebel JG, Van Winkle TJ, Goldstein RE, Serfilippi LM (2008) Clinicopathologic, histologic, and toxicologic findings in 70 cats in advertently exposed to pet food contaminated with melamine and cyanuric acid. J Am Vet Med Assoc 233:729–737
139. Puschner B, Poppenga RH, Lowenstine LJ, Filigenzi MS, Pesavento PA (2007) Assessment of melamine and cyanuric acid toxicity in cats. J Vet Diagn Invest 19:616–624
140. WHO (2009) Toxicological and health aspects of melamine and cyanuric acid. World Health Organization, Geneva. Available: http://www.who.int/foodsafety/fs_management/infosan_events/en/index.html. Accessed 21 Oct 2009
141. WHO (2008) Melamine and cyanuric acid: toxicity, preliminary risk assessment and guidance on levels in food. World Health Organization, Geneva. Available: http://www.who.int/entity/foodsafety/fs_management/Melamine.pdf. Accessed 21 Oct 2009)
142. WHO (2004) Sodium dichloroisocyanurate. In: Safety evaluation of certain food additives and contaminants, vol 52, WHO Food Additives Series. World Health Organization, Geneva
143. Heck HD, Tyl RW (1985) The induction of bladder stones by terephthalic acid, dimethyl terephthalate, and melamine (2, 4, 6-triamino-s-triazine) and its relevance to risk assessment. Regul Toxicol Pharmacol 5:294–313
144. NTP (1983) Carcinogenesis bioassay of melamine (CAS No. 108-78-1) in F344/N Rats and B6C3F1 Mice (Feed Study). TR 245. Research triangle park, NC: National toxicology program. Available: http://ntp.niehs.nih.gov/ntp/htdocs/LT_rpts/tr245.pdf. Accessed 21 Oct 2009
145. Reimschuessel R, Gieseker CM, Miller RA, Ward J, Boehmer J, Rummel N, Heller DN, Nochetto C, Hemakanthi de Alwis GK, Bataller N, Andersen WC, Turnipseed SB, Karbiwnyk CM,

Satzger RD, Crowe JB, Wilber NR, Reinhard MK, Roberts JF, Witkowski MR (2008) Evaluation of the renal effects of experimental feeding of melamine and cyanuric acid to fish and pigs. Am J Vet Res 69(9):1217–1228
146. apek P, Dickerson TJ (2010) Sensing the deadliest toxin: technologies for botulinum neurotoxin detection. Toxins 2:24–53
147. Singh NP, Anuradha S, Dhanwal DK, Singh K, Prakash A, Madau K, Agarwal SK (2000) Epidemic dropsy – a clinical study of the Delhi outbreak. J Assoc Physicians India 48:877–880
148. Arnon SS, Schechter R, Inglesby TV, Henderson DA, Bartlett JG, Ascher MS, Eitzen E, Fine AD, Hauer J, Layton M, Lillibridge S, Osterholm MT, O'Toole T, Parker G, Perl TM, Russell PK, Swerdlow DL, Tonat K, Working Group on Civilian Biodefense (2001) Botulinum toxin as a biological weapon: medical and public health management. JAMA 285(8):1059–1070
149. Wagacha JM, Muthomi JW (2008) Mycotoxin problem in Africa: current status, implications to food safety and health and possible management strategies. Int J Food Microbiol 124:1–12
150. Kumar V, Basu MS, Rajendran TP (2008) Mycotoxin research and mycoflora in some commercially important agricultural commodities. Crop Protec 27:891–905
151. Refai MK (1988) Aflatoxins and aflatoxicosis. J Egypt Vet Med Assoc 48:1–19
152. Wary BB (1981) Aflatoxin, hepatitis B-virus and hepatocellular carcinoma. N Engl J Med 305:833–843
153. Sedmikova M, Reisnerova H, Dufkova Z, Barta I, Jilek F (2001) Potential hazard of simultaneous occurrence of aflatoxin B1 and Ochratoxin A. Vet Med Czech 46:169–174
154. Paterson RRM (2006) Fungi and fungal toxins as weapons. Mycol Res 10(9):1003–1010
155. Bennett JW, Klich M (2003) Mycotoxins. Clin Microbiol Rev 16(3):497–516
156. He J, Zhou T, Young JC, Boland GJ, Scott PM (2009) Chemical and biological transformations for detoxification of trichothecene mycotoxins in human and animal food chains: a review. Trends Food Sci Technol 21(2):67–76
157. Rocha O, Ansari K, Doohan FM (2005) Effects of trichothecene mycotoxins on eukaryotic cells: a review. Food Addit Contam 22:369–378
158. Lancova K, Hajslova J, Poustka J, Krplova A, Zachariasova M, Dostalek P, Saschambula L (2008) Transfer of Fusarium mycotoxins and "masked" deoxynivalenol (deoxynivalenol-3-glucoside) from field barley through malt to beer. Food Addit Contam 25:732–744
159. Pittet A (2001) Natural occurrence of mycotoxins in foods and feeds: a decade in review. In: De Koe WJ, Samson RA, van Egmond HP, Gilbert J, Sabino M (eds) Mycotoxins and phycotoxins in perspective at the turn of the Millennium. W.J. de Koe, Wageningen, pp 153–172
160. Goyarts T, Dnicke S, Valenta H, Ueberschär KH (2007) Carry-over of Fusarium toxins (deoxynivalenol and zearalenone) from naturally contaminated wheat to pigs. Food Addit Contam 24:369–380
161. Seeling K, Dänicke S, Valenta H, Van Egmond HP, Schothorst RC, Jekel AA, Lebzien P, Schollenberger M, Razzazi-Fazeli E, Flachowsky G (2006) Effects of Fusarium toxin-contaminated wheat and feed intake level on the biotransformation and carry-over of deoxynivalenol in dairy cows. Food Addit Contam 23:1008–1020
162. Valenta H, Dänicke S (2005) Study on the transmission of deoxynivalenol and deepoxy-deoxynivalenol into eggs of laying hens using a high-performance liquid chromatography-ultraviolet method with clean-up by immunoaffinity columns. Mol Nutr Food Res 49:779–785
163. Kortepeter M, Christopher G, Cieslak T, Culpepper R, Darling R, Pavlin J, Rowe J, McKee K Jr, Eitzen E Jr (eds) (2001) USAMRIID's medical management of biological casualties handbook. US Army Medical Research Institute of Infectious Diseases, Fort Detrick
164. Bhaskar ASB, Deb U, kumar O, Rao PVL (2008) Abrin induced oxidative stress mediated DNA damage in human leukemic cells and its reversal by N-acetylcycteine. Toxicol In Vitro 22:1902–1908
165. Sekheta MAF, Sahtout AH, Farid N, Sekheta NF, Pantovic N, Al Omari AT (2006) Terrorist threats to food & water supplies and the role of HACCP implementation as one of the major effective and preventive measures. Internet J Food Saf 8:30–34

166. Sobel J, Khan AS, Swerdlow DL (2002) Threat of a biological terrorist attack on the US food supply: the CDC prospective. Lancet 359:874–880
167. Sekheta MAF, Sahtout AH, Sekheta NF, Kapkovic M, Pantovic N (2005) The HACCP implementation and the mental illness of food handlers as the 4th eventual hazard. Internet J Food Saf 6:5–10
168. Miraglia M, Marvin HJP, Kleter GA, Battilani P, Brera C, Coni E, Cubadda F, Mumtaz MM, Ruiz P, De Rosa CT (2007) Toxicity assessment of unintentional exposure to multiple chemicals. Toxicol Appl Pharmacol 223:104–113
169. Azzazy HME, Mansour MMH (2009) In vitro diagnostic prospects of nanoparticles. Clin Chim Acta 403(1–2):1–8
170. IPCC (2007) Intergovernmental panel on climate change – IPCC. Mitigation from a cross-sectorial perspective. Working group III report, mitigation of climate change – Chapter 11. Available in: http://www.ipcc.ch. Accessed 20 Sept 2007
171. FAO (1998) Food and Agriculture Organization of the United Nations. Declaração de Roma sobre a segurança alimentar mundial e plano de acção da cimeira mundial da alimentação. Available in: http://www.fao.org/docrep/003/w3613p/w3613p00.htm. Accessed: 05/07/2007)
172. United Nations (2007) World population prospects: the 2006 revision and world urbanization prospects. Available in: http://esa.un.org/unpp/p2k0data.asp. Accessed 25 May 2007
173. WWF (2007) WWF – position on biofuels in the EU; 2007. Available in: http://www.wwf.org. Accessed 15 May 2007
174. WHO (2006) Total diet studies: a Recipe for safer food. International Food Safety Authorities Network (INFOSAN). INFOSAN information note No. 06/2006 –Total diet studies; 3 pp
175. Kleter G, Prandini A, Filippi L, Marvin HJP (2009) Identification of potentially emerging food safety issues by analysis of reports published by the European Community's Rapid Alert System for Food and Feed (RASFF) during a four-year period. Food Chem Toxicol 47(5):932–950
176. De Rosa C, Hicks T, Ashizawa HE, Pohl A, Mumtaz M (2006) A regional approach to assess the impact of living in a chemical world. Ann NY Acad Sci 1076:829–838
177. Department of Health and Human Services – Centers for Disease Control and Prevention (2005). Third National Report on Human Exposure to Environmental Chemicals. www.clu-in.org/download/contaminantfocus/pcb/third-report.pdf
178. Durocher J (2003) Better safe. Restaur Bus 102(8):90–91
179. Haynes JR, Beck JE (2004) Evaluating security readiness in foodservice using (SARA), safety analysis risk assessment. J Fam Consum Sci Educ 22(1):1, Spring/Summer 2004
180. Pisante M, Piva G, Prandini A, Toti L, van den Born GJ, Vespermann A (2009) Climate change and food safety: an emerging issue with special focus on Europe. Food Chem Toxicol 47(5):1009–1021
181. Prewitt M (May 24, 2004) The 7 cardinal sins of foodservice: failure to make safe haven for staff risks harm to bottom line. Nations Restaurant News, pp 62–64

Chapter 7
Defending the Safety of the Global Food System from Intentional Contamination in a Changing Market

Francis (Frank) F. Busta and Shaun P. Kennedy

Abstract Intentional contamination of the food supply poses a real and potentially catastrophic threat to society. Overall, it has the potential to result in disastrous and far-reaching effects, including direct morbidity and/or mortality, disruption of food distribution, loss of consumer confidence in government and the food supply, business failures, trade restrictions, and ripple effects on the economy. Key interrelated factors specific to food and the food system create this unusual vulnerability, both structural and social. The efficiency of the food system enables products derived from a wide range of global sources to be sourced, produced, and distributed rapidly due to the speed of national and global just-in-time supply chains. The food industry's routine food safety measures are not designed to protect against high-impact deliberate contamination. When contamination occurs, identification of its nature and extent may take days, weeks, or even longer. Unintentional foodborne illness can further complicate recognition of intentional contamination events due to the delay in positive association of illnesses to the intentional event.

The food/agriculture sector's infrastructure must be strengthened to mitigate potential harm resulting from deliberate contamination, thereby making the food system less vulnerable to attack or destructive economic outcomes.

New upward price pressures, declining economies, and constantly changing global trade along the food system supply chain have introduced a new urgency for greater diligence in food defense against deliberate contamination with either economic or terrorist motives. Initiatives include the development of specific countermeasures to minimize or eliminate vulnerabilities, as well as the development of practical solutions that enhance the capability to rapidly identify, contain, respond to, and recover from intentional contamination, both real and threatened. These activities must encompass the entire worldwide farm-to-table food system, from pre-farm inputs through retail sale, consumer food consumption, and public health system response.

F.F. Busta (✉) and S.P. Kennedy
National Center for Food Protection and Defense, University of Minnesota, St. Paul, MN 55108, USA
e-mail: fbusta@umn.edu; kenne108@umn.edu

M. Hefnawy (ed.), *Advances in Food Protection*, NATO Science for Peace and Security Series A: Chemistry and Biology 1, DOI 10.1007/978-94-007-1100-6_7,

Keywords Food safety • Intentional contamination

7.1 Introduction

Intentional contamination of the food supply poses a real and potentially catastrophic threat to society. Overall, it has the potential to result in disastrous and far-reaching effects, including direct morbidity and/or mortality, disruption of food distribution, loss of consumer confidence in government and the food supply, business failures, trade restrictions, and dramatic ripple effects on the economy. This chapter will address relationships of food protection, food quality, food safety, food defense, and food security. It will highlight vulnerabilities and threats to the entire food supply chain from agricultural inputs to the consumer, including occurrences of intentional contamination. The complexity of the global food supply chain and the relative ease of access for deliberate addition of undocumented materials will point to the challenges of control in current times of economic stress and suggest responses with some food defense actions.

Many questions can be asked in regard to intentional contamination of food in a climate of global unrest, economic stress, natural and industrial disasters, and worldwide change. Is there general, corporate, governmental, or public awareness of the reasons for concern about intentional contamination of the food system? How do food safety, food defense, and food protection relate to each other? How does food defense relate to economically motivated adulteration? Are we doing what needs to be done globally? How can academic research support food defense and food safety throughout the entire food system?

The following will assist to describe and define the food continuum paradigm:

- Food security: supply sufficiency – access by all people to enough readily available, nutritionally adequate and safe food for an active, healthy life. This includes an ability to acquire acceptable foods in a socially acceptable way.
- Food safety: system reliability – reducing exposure to naturally-occurring hazards, errors, or failures. This relates to well-recognized, regularly and routinely experienced, unintentional occurrences.
- Food defense: system resiliency – reducing the impact of attacks on the food system. This relates to intentional or deliberate introduction of materials into the food supply to cause physical or economic harm. It is criminal, asymmetric, and sporadic.
- Food protection: safety and defense "umbrella" – overall safety and defense of the food supply system that also weighs heavily on food quality.
- Food quality: conformance to specifications and standards – dependent upon food safety that is in turn dependent upon food defense.
- Economically motivated adulteration (EMA): fraudulent addition of non-authentic substances for financial advantage with the goal of not being detected (see Fig. 7.1).

Threats to the food system from contamination come in many and various forms, be they unintentional or deliberate [1, 2]. Some are naturally occurring and must be dealt

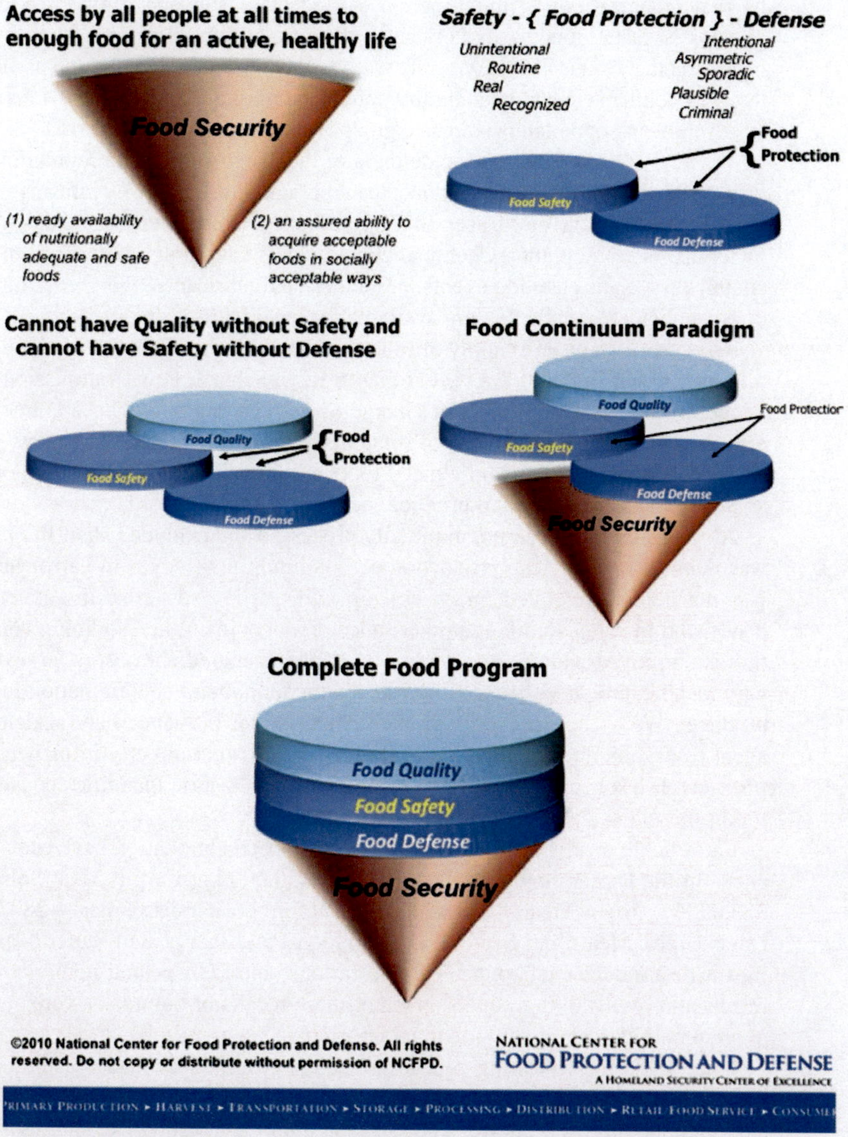

Fig. 7.1 Relationships in complete food program

with on a routine basis. Some develop due to negligence or accidental contamination of the food supply somewhere along the supply chain. The more challenging and nefarious contamination stems from deliberate introduction of threat agents of some type. Disgruntled employees may wish to do harm to their employers or the organization at which they had worked. Violent activist groups may use this means to emphasize their cause, either against the food system they are attacking or some system represented by that organization. Criminal action can occur as subversive extortion of some component of the food system or can be the introduction of an undeclared substitute ingredient that results in economically motivated adulteration. Finally, and probably the threat of deliberate contamination that causes the most anxiety, is that of international or government-supported or -directed groups or individuals – the terrorists.

Whether it is inadvertent or deliberate, the range of possible contaminants is broad [3]. Biological contaminants can be infectious microorganisms such as pathogenic bacteria or viruses, toxins extracted from microorganisms or plant materials, or toxic plants. Chemical contaminants can be agricultural or industrial chemicals such as cleaning agents, pesticides, and lubricants, allergens, or a variety of poisonous chemicals that are available for terrorist use [4].

The cost of an unsafe supply of food due to foodborne illness is shocking – $152 billion per year in the U.S.A. according to R. L. Scharff, Pew Commission Report of the Produce Safety Project at Georgetown University [5]. These extraordinary estimates prompt a brief review of some of the emerging recent food safety threats. There have been recent unanticipated food and infectious agent combinations such as *Salmonella* Saintpaul and produce (peppers or tomatoes or both).

Abuse of underprocessed, minimally processed and extended shelf life products was demonstrated by the production of botulinum neurotoxin in carrot juice that was not kept refrigerated, in cheese dip with garlic and herbs, in canned green beans, and in canned chili and other sauce products [6]. When sourcing consolidation is employed and the broad use of a common ingredient occurs, an extraordinary recall can be necessary, as was seen with *Salmonella* contamination in peanut products. We are experiencing climate changes that influence natural deleterious agent reservoirs and produce harmful algal blooms that directly influence fishery products. It has become readily apparent that globalization magnifies conventional food safety considerations and potential impacts.

Large scale recalls and outbreaks as a result of contamination have been experienced in the U.S.A., e.g., Fairbank Farms – 250,000 kg; Cargill Meat Solutions – 385,000 kg; Topps Meats – 10 million kg; Sanford green mussel meat – 280,000 kg; Davis Creek Meats and Seafood – 59,000 kg; Peter Pan peanut butter – over 600 ill; Castle Produce cantaloupes – 2,560 cartons; and PCA peanut items – over 500 recalls and over 4,000 products. None of these recalls or outbreaks were attributed to an intentional act by an individual or group.

Since 1998 the following agents have been identified as involved in deliberate contamination events: arsenic, cyanide, feces, herbicide, household cleaner, insecticide, nicotine sulfate, pesticide, rat poison, tetramine, and thallium. Countries that have experienced intentional contamination events include Australia, Canada, China/Hong Kong, Iraq, Italy, Japan, Korea, Philippines, Taiwan, Thailand, and the U.S.A. [7].

Intentional food contamination is not limited to recent times. Historically, the military utilized this as one of their weapons against troops and civilians [8]. The earliest well-documented use of food as a military weapon was in 590 B.C. in the war between Athens and Kirrha of the Amphictyonic League. The Athenians poisoned stored water with helleborous root allowing them to overcome the city with gastrointestinal illness. The Carthaginian general Maharbal used wine contaminated with mandragora in battling tribesmen, and Julius Caeser in 75 B.C. in Miletus slaughtered Cilician pirate captors by lacing wine with the same mandrake poison delivered to the pirates with their ransom [9]. During World War II the Japanese Army experimented with bacterial pathogens including *B. anthracis, Shigella* spp., *Vibrio cholerae*, and *Yersinia pestis* introduced into candy and dropped from planes over villages as well as with vegetables, meat, and fish [10, 11]. The most well-known recent large-scale intentional contamination event in the U.S.A. occurred in 1987 with the Rajneeshee cult in Oregon [12]. The cult was testing a plan to influence a local election by contaminating salad bars with *Salmonella* Typhimurium, resulting in 751 confirmed cases. All of this was revealed by a member of the cult over a year later in another investigation. Although many events are not widely publicized, a significantly large number of intentional contaminations of food have been documented by Mohtadi and Murshid [13]. The following list is a series of examples of intentional food contamination events in one country, China, over a 5-year period:

1997: Insecticide contaminated pork provided to villagers by another resident (44 ill)
1999: Nitric acid contamination of donkey soup by competitors to the food vendor (148 ill)
1999: Rat poison contamination of sweet rolls by competitors to the food vendor (48 ill)
2001: Rat poison contaminated noodles by a competitor to the food vendor (120 ill)
2002: Rat poison contaminated school lunch (92 ill)
2002: Rat poison contaminated breakfast foods by a competitor to the food vendor (400 ill, 41 dead)

More recently state media reported that a man was arrested in China accused of poisoning gyoza dumplings in a case that led to a diplomatic rift with Japan. Temporary factory worker, Lu Yueting, 36, allegedly put insecticide in some frozen dumplings because he was unhappy with his pay and colleagues. The food was exported to Japan, where ten people became ill, sparking a scare over Chinese food. After much diplomatic effort and bilateral suspicion, officials signed an agreement to improve food safety standards and restore confidence [14].

Emerging food defense threats stemming from intentional contamination span a broad spectrum of individuals and organizations. The transition of the agriculture sectors in the developing world from subsistence agriculture to agribusiness with more effort directed at commodity and value-added exports has dramatically increased access to the food supply chain and consequently increased vulnerabilities. These threats may originate with disgruntled employees who wish to disrupt the organization's production for a variety of reasons – usually these actions are aimed at economic disruption and not focused on high mortality or morbidity. There are occasions where a criminal element attempts extortion with the threat of intentionally contaminating specific food products. Similar actions have been generated by psychologically disturbed individuals or extremist special interest groups who wish to stop a specific type of food supply chain activity. They might focus on production of a specific crop, the slaughter of food animals, the use of a type of transport, or the general consumption of some item. Of course, since the terrorist events of September 11, 2001, internationally sponsored or funded individual terrorists or groups have received extensive consideration in defending the safety of the food system.

The recent global economic crisis has exacerbated problems that arise from producers or others along the supply chain introducing alternative materials without declaring the action. Obviously, greed is a major motivation, and ignorance may generate major problems. Food fraud is not a new problem [15]. Compared to many intentional contaminators that wish to be discovered for the "political" statement, these economically motivated adulterators intend never to be discovered. Unfortunately when public or animal health suffers, the morbidity or mortality leads to discovery. The Rapid Alert System for Food and Feed listed the following numbers of RASFF Fraud Reports in 2003–2006:

Animal feed	7	Meat	58
Bakery	3	Not specified	8
Beverage	1	Nut	11
Composite, mixed	42	Oil and fat	4
Confectionery	4	Other	6
Dairy	11	Poultry	20
Egg product	2	Seafood	46
Food supplement	3	Snack	1
Fruit, processed fruit	1	Spice, condiment	13
Honey, bee product	5	Vegbl./cereal/other	2

Another example of the problem is the 2006–2007 melamine in gluten adulteration [16]. This resulted in major cat and dog illnesses and subsequent large-scale pet food recalls due to melamine presence in gluten from China sold through a complex supply chain to manufacturers of pet food produced in the U.S.A. (see Fig. 7.2). These economically motivated adulterations present the potential for unanticipated or unusual threat agents such as melamine, *Salmonella* Typhimurium, etc. It is estimated by the Grocery Manufacturers Association [17] to result in losses of 2–15% of sales to the impacted firm.

There are many examples of this problem. A recent situation in China was described by Michelle Greenhalgh in *Food Safety News*; March 22, [18] when Chinese cooking oil was found contaminated and emphasized the problem. She related the following citations from news items in China. The State Food and Drug Administration, China's food safety watchdog, ordered a nationwide inspection of all cooking oil as reports released by *The China Daily* indicated that almost one-tenth of Chinese supplies contained cancer-causing agents and were made illegally. *The Taipei Times* reported that one-tenth of drainage oil is most likely made from highly toxic, recycled kitchen or restaurant waste containing aflatoxin. Chronic exposure to aflatoxin can cause cancer. He Dongping, a food science expert at the Wuhan Polytechnic University, told the *China Youth Daily* "People in China consume about 2–3 million tons of illegal cooking oil every year." Dongping cited

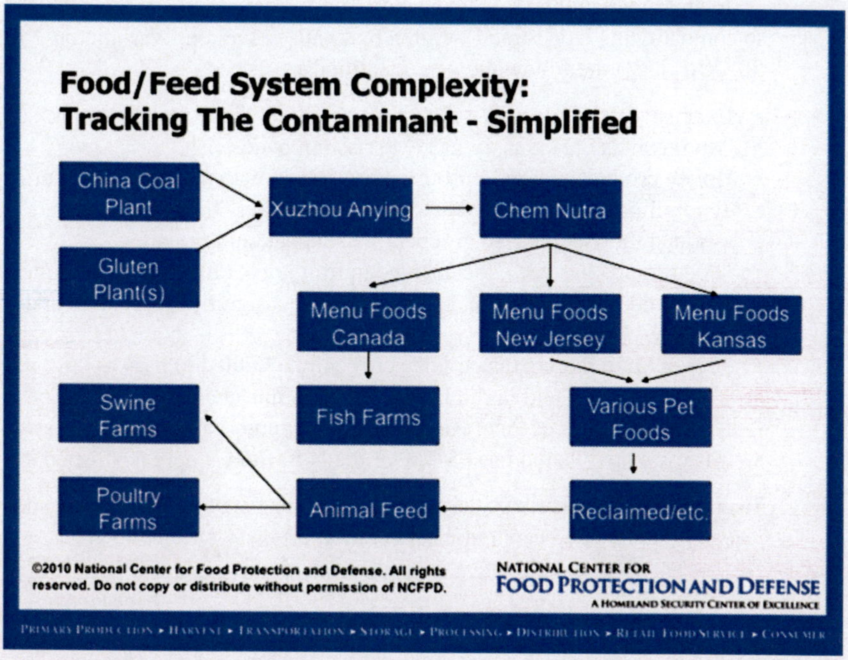

Fig. 7.2 Simplified flow chart of supply chain for gluten adulterated with melamine destined for pet food, demonstrating the complexity of the food and feed system and the subsequent difficulty in trace back, track forward of a contaminant

statistics indicating that the Chinese population consumes about 22.5 million tons of cooking oil per year. This discovery comes on the heels of ongoing melamine-tainted milk trials and other violations, putting China in the food safety spotlight. It would be remiss not to recognize that economically motivated adulteration of oils occurs in places other than China; e.g., deliberately contaminated sunflower oil in Europe [19, 20].

The presence of an unsafe ingredient in some flours was reported by Wang Yan [21] in the *China Daily on* April 8, 2010. Recent investigations found that some bleaching agents widely used in flour production contain as much as 30% pulverized lime, a substance that has been linked to health problems. Normal price of bleaching agent is 11,000 yuan ($1,611) per ton, but from the Yuzhong company only 9,000 yuan per ton. Such news did not surprise some within the industry. "Their low prices resulted from adding pulverized lime," which is not only cheaper but also weighs more. "I don't think it's a hidden rule in the industry," he said, adding problems caused by unregulated use of additives in food production is unavoidable in the current developmental stage of the society. "Only when the number of peasant households in the country is reduced, say from the current 200 million to 2 million, could we possibly solve the food-safety problem at the source," he said.

In an article on the U .S. Food and Drug Administration (FDA) being pressured to combat rising 'food fraud' written by Lyndsey Layton, Washington Post (March 30, 2010 [22]) the following were identified:

- Expensive "sheep's milk" cheese was really made from cow's milk.
- "Sturgeon caviar" was, in fact, Mississippi paddlefish.
- Honey producers diluted "100% pure" honey with sugar beets or corn syrup.
- Ten million pounds of cheap, frozen catfish fillets from Vietnam were identified as much more expensive grouper, red snapper, and flounder.
- "Food fraud" has been documented in fruit juice, olive oil, spices, vinegar, wine, spirits and maple syrup, and appears to pose a significant problem in the seafood industry.
- 50% of "pure pomegranate juice concentrate" failed to pass as only juice.
- 77% of snapper sold in the United States is mislabeled.
- 18 million bottles of Pinot Noir were made from cheaper merlot and syrah grapes
- It is estimated that at least 5–7% of the U.S. food supply is affected.

To emphasize the current status, the following is a list of foods that were identified in recently documented economic adulteration events:

Pet food	Maple syrup	Olive oil
Vegetable proteins	Honey	Sunflower oil
Cooking oils	Infant formula	Seafood
Mushrooms	Dairy products	Cheese
Apple juice	Eggs	Horseradish
Orange juice	Gums	Tofu
Grape juice	Honey	
Grapefruit juice	Vanilla extract	

There were numerous practices that assist in this fraud, including trans-shipping to avoid duties or inspections, port shopping to avoid inspections, diverted products entering commerce for grey/black market, and general economic adulteration for potential gain or to maintain revenue. It must be emphasized at this point that if these situations currently are so widespread, it is not difficult to imagine how intentional contamination for the purpose of doing harm could easily occur.

The comprehensive food supply system from agricultural inputs to consumers is a complex supply chain with intermediate complex operations. Consequently, there are multiple vulnerabilities available to attack. The complexity is demonstrated in Fig. 7.3 with a simplified supply chain for the standard American cheeseburger, in Fig. 7.4 with a list of possible ingredients in each component of the cheeseburger considering that each of the ingredients in turn has its own supply chain, and in Fig. 7.5 with the possible global sources of some of the ingredients.

Food systems present numerous opportunities for contaminant access. Frequently there is a lack of a treatment sufficiently adequate for inactivation or neutralization of a contaminant after its addition. Many foods are prepared in large volumes or batches that are well mixed for best homogeneity. Some foods are ideal for the amplification of a contaminant if it is capable of increasing in concentration by either growth or partitioning. Many foods are broadly distributed to consumers who in turn consume

Fig. 7.3 Simplified supply chain for the standard American cheeseburger

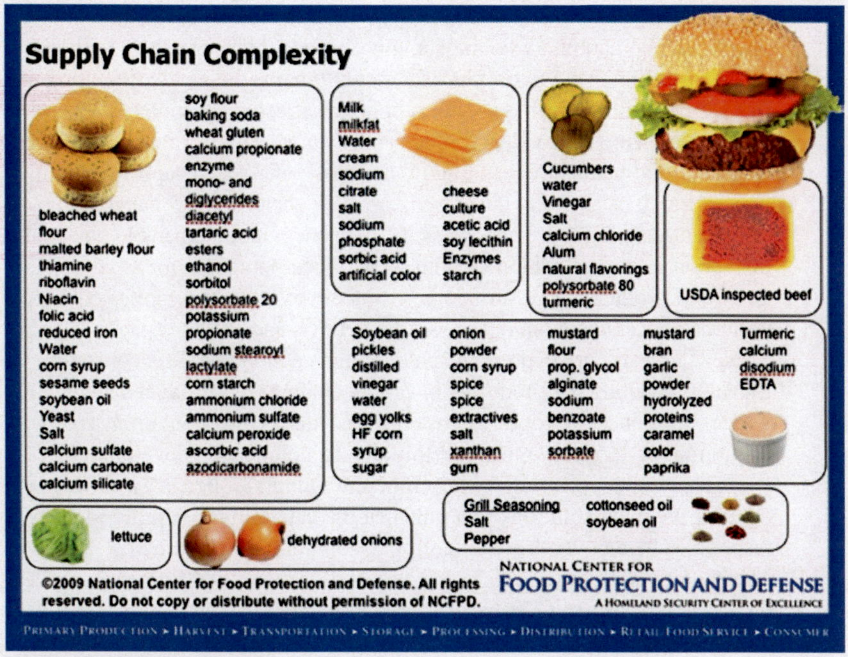

Fig. 7.4 List of possible ingredients in each component of the cheeseburger

Fig. 7.5 Possible global sources of some of the cheeseburger ingredients

them soon after purchase. In addition, some foods are prepared for and distributed to "high risk populations" who may be more susceptible or sensitive to the contaminant or in less strong to survive a physical insult. Finally, the slow recognition of an event or outbreak will result in a greater number of affected consumers.

Coupled with these vulnerabilities is the extraordinary complexity of the food and agriculture infrastructure resulting in the most complicated supply chain that exists in any sector. It is globally dispersed, privately held, highly integrated, flexible yet redundant with limited excess capacity, dynamic with innumerable potential points of disruption or contamination or both. It has been optimized for rapid delivery of low cost food products from all sources. Since it is such a complicated supply chain, perturbations at one point have cascading effects and excess capacity is very limited.

New upward price pressures, declining economies, and constantly changing global trade along the food system supply chain have introduced a new urgency for greater diligence in food defense against deliberate contamination with either economic or terrorist motives. Initiatives include the development of specific countermeasures to minimize or eliminate vulnerabilities, as well as the development of practical solutions that enhance the capability to rapidly identify, contain, respond, and recover from intentional contamination, both real and threatened. The National Center for Food Protection and Defense (NCFPD), a U.S. Department of Homeland Security Center of Excellence, based at the University of Minnesota, was launched in 2004 to address these needs. A multidisciplinary and action-oriented research consortium, NCFPD addresses the vulnerability of the nation's food system to attack through intentional contamination with biological or chemical agents. NCFPD's research and education program is aimed at reducing the potential for contamination at any point along the food supply chain and mitigating potentially catastrophic public health and economic effects of such attacks. The program incorporates cutting-edge research across a wide range of disciplines, taking a comprehensive, farm-to-table view of the food system and encompassing all aspects from primary production through transportation and food processing to retail and food service.

The vision of NCFPD is direct and forthright: *Defending the safety of the food system through research and education.* To accomplish this mission, NCFPD has partnered with a broad range of academic institutions since its inception (Fig. 7.6). Other partnerships have also developed with the private sector and NGOs as they assist NCFPD in planning, evaluating, and implementing the research and education program (Fig. 7.7). The third component of NCFPD's relationships is with the public sector including government agencies at the international, national, state, and local level (Fig. 7.8). The diversity and breadth of these collaborations have permitted NCFPD to pursue its vision. The geographical distribution of collaborators in 2010 emphasizes the global distributions of our diverse collaborators

Fig. 7.6 Logos of academic institutions that have partnered with NCFPD since 2004

Fig. 7.7 Logos of private sector industries and NGOs that have collaborated with NCFPD since 2004

Fig. 7.8 Logos of government agencies and organizations that have collaborated with NCFPD since 2004

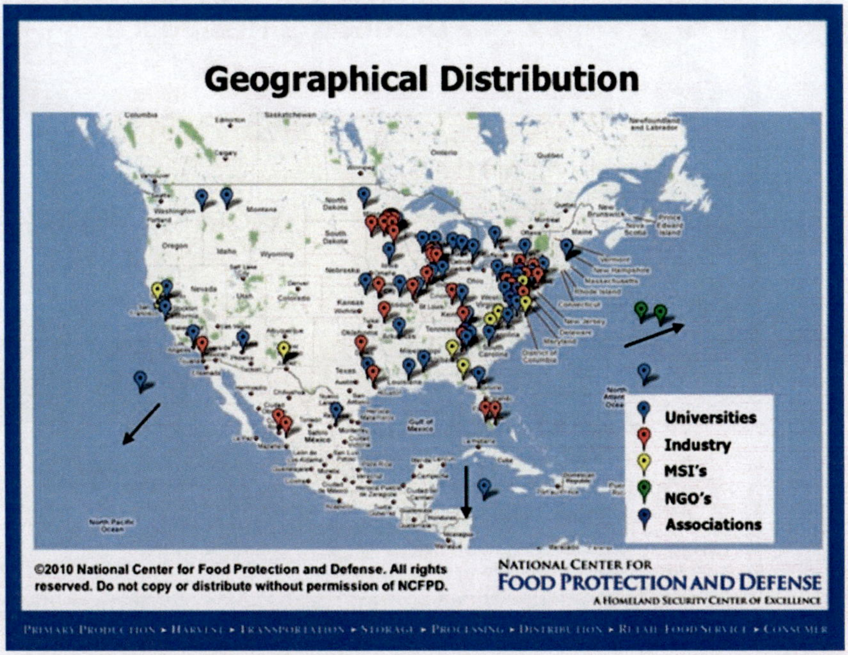

Fig. 7.9 Geographic distribution of partners and collaborating institutions and agencies including universities, industries, minority-serving institutions, non-government organizations, and various member associations

and partners and emphasizes our consideration of the global impact on food defense (Fig. 7.9). An example that emphasizes the global efforts of NCFPD is the table-top exercise in food defense, identified as "Demeter's Resilience," in 2008, assessing global capabilities of response to an intentional contamination event (Fig. 7.10 [23]).

The goal of NCFPD is to reduce, through research and education, the potential for catastrophic food system events by:

- Rendering targets unattractive to individuals wishing to do harm.
- Rapidly delivering effective recovery efforts should an event occur.
- Rapidly and accurately detecting attacks before, during, and after an occurrence.
- Responding effectively to minimize consequences of an event.
- Training new scientists and practicing professionals to understand food defense.
- Partnering and collaborating to ensure success of the mission of NCFPD.

Fig. 7.10 Logos of academic and government institutions that participated in the table-top exercise in food defense, identified as "Demeter's Resilience," in 2008, assessing global capabilities of response to an intentional contamination event

NCFPD has four primary research themes that are also coupled with education:

System Strategies, Event Modeling, Agent Behavior (chemical and biological), and Risk and Crisis Communications.

- "System Strategies" involves tools for the public and private sector to assess and improve their capabilities in food system event prevention, response and recovery. It also is developing evaluation tools for stakeholders to optimize investments for mitigation of food system events. The publication by [24] presents an example of the research in this theme.
- "Event Modeling" includes realistic, stakeholder engaged, data driven models of potential food system events for (1) risk, vulnerability, and consequence assessment, (2) intervention/countermeasure effectiveness assessment, (3) decision support tools for early event detection and rapid response, and (4) education and awareness training. Research published by Hartnett et al. [25] serves as an example of the research in this theme.
- "Agent Behavior" covers fundamental understanding of food-agent interactions to enable rapid detection and removal of select agents from complex sample matrices, process-based inactivation, post-event facility-compatible decontamination, and safe contaminated-product disposal. This addresses both biological agents such as *Bacillus anthracis* spores and ricin as well as chemical agents

such as nicotine sulphate and fluoroacetic acid. Publications by Park et al. [26] and Hilgren et al. [27] describe research that serve as examples for this theme.
- "Risk and Crisis Communications" focuses on best practices in food defense crisis situations where planning before, acting during, and following up after the event are key to success. This is fundamental strategic planning, proactive delivery, and effective response. Venette et al. [28] published findings that reflect research in this theme.

Finally, it is essential that educational opportunities are identified and information on developing food defense plans be listed. This list identifies tools that are available for effective program development.

First the reader is reminded that the NCFPD website http://www.ncfpd.umn.edu has much information and essential links for food defense research and education.

The FDA Food Defense Oversight Team offers updated tools and resources available to the public, including improvements in the CARVER software, the FDA online mitigation database and food-related emergency exercises (scenarios that can be adapted to meet various users' needs). Additional resources are found on the their homepage http://www.FDA.gov/FoodDefense or at http://www.fda.gov/Food/FoodDefense/default.htm

FDA works with other government agencies and private sector organizations to help reduce the risk of tampering or other malicious, criminal, or terrorist actions on the food and cosmetic supply.

USDA Food Safety and Inspection Service (FSIS) offers through Food Defense and Emergency Response the New General Food Defense Plan: http://www.fsis.usda.gov/food_defense_&_emergency_response/index.asp A new voluntary Food Defense Plan also is now available for small and very small plants at Podcasts on Food Defense; Meat, Poultry and Egg Product Inspection.

At the international level, the World Health Organization (WHO) addresses terrorist threats to the food system:

http:// www.who.int/foodsafety/publications/general/en/terrorist.pdf

The Food and Agriculture Organization (FAO) has recently developed EMPRES for assistance in food protection with support from NCFPD:

http:// www.fao.org/ag/againfo/programmes/en/empres/home.asp

7.2 Summary and Conclusions

Intentional contamination of the food supply poses a real and potentially catastrophic threat to society. A stressed economy with upward price pressures and a constantly changing global trade along the food system supply chain have introduced a new urgency for greater diligence in food defense against deliberate contamination with either economic or terrorist motives. The global food/agriculture sector's infrastructure must be strengthened to mitigate potential harm resulting from deliberate contamination. Research initiatives are needed to address each of these concerns. Educational efforts are needed in food supply chain management, risk communication, and food defense programs related to counter-terrorism and counter-economic-adulteration. Awareness and cooperation among the public and private sector members along the entire food supply chain are required to minimize vulnerabilities and enhance overall food protection.

Defending the safety of the food system through research and education

Acknowledgement This material is based upon work supported by the U.S. Department of Homeland Security under Grand Award Number 2007-ST-061-000003-02.

Disclaimer. The views and conclusions contained in this document are those of the authors and should not be interpreted as necessarily representing the official policies, either expressed or implied, of the U.S. Department of Homeland Security.

References

1. Kinsey JD, Stinson TF, Degeneffe DJ, Ghosh K, Busta FF (2009) Consumers response to a new food safety issue: food terrorism. In: Barbosa-Canovas GV (ed) Global issues in food science and technology. Elsevier, New York, pp 145–161
2. Ostrowsky J, Kennedy SP, Busta FF (2006) Priorities for research in food defense. In: Addressing foodborne threats to health: policies, practice, and global coordination, Forum on Microbial Threats, Session VII, NAS Institute of Medicine, Report. The National Academies, Washington, DC, pp 247–254
3. Kennedy SP, Busta FF (2007) Biosecurity: food protection and defense. In: Doyle MP, Beuchat LR (eds) Food microbiology: fundamentals and frontiers, 3rd edn. ASM, Washington, DC, pp 87–102, Chapter 5
4. Kuiper HA, Kleter GA (2009) Special issue: early awareness of emerging risks to food and feed safety. Food Chem Toxicol 47(5):909–1039
5. Scharff RL (2010) Health-related costs from foodborne illness in the United States. Pew Commission Report of the Produce Safety Project at Georgetown University. www.producesafetyproject.org. 3 Mar 2010
6. Center for Science in the Public Interest (2009) Outbreak alert!: analyzing foodborne outbreaks 1998 to 2007, 11th edn. Center for Science in the Public Interest, Washington, DC
7. Dalziel GR (2009) Food defense incidents 1950–2008: a chronology and analysis of incidents involving the malicious contamination of the food supply chain. Center of Excellence for National Security, S. Rajaratnam School of International Studies, Singapore
8. Poupard JA, Miller LA (1992) History of biological warfare: catapults to capsomeres. Ann NY Acad Sci 666:9–20
9. Mayor A (2004) Poison waters, deadly vapor and sweet sabotage. In: Greek fire, poison arrows & scorpion bombs: biological and chemical warfare in the ancient world. Overlook Press/Peter Mayer Publishers, Woodstock, pp 99–118, pp 145–169
10. Harris S (1992) Japanese biological warfare research on humans: a case study of microbiology and ethics. Ann NY Acad Sci 666:21–52
11. Harris SH (2003) Japanese biomedical experimentation during the World-War-II Era. In: Lounsbury DE, Bellamy RF (eds) Military medical ethics, vol II. Office of the Surgeon General, Department of the Army, Washington, DC, pp 463–506
12. Torok TJ, Tauxe RV, Wise RP, Livengood JR, Sokolow R, Mauvais S, Birkness KA, Skeels MR, Horan JM, Foster LR (1997) A large community outbreak of salmonellosis caused by intentional contamination of restaurant salad bars. JAMA 278:389–395

13. Mohtadi H, Murshid AP (2009) Risk of catastrophic terrorism: an extreme value approach. J Appl Econometrics 24(4):537–559
14. Food Production Daily (2010) Japan and China sign food safety pact. http://www.foodproductiondaily.com/Quality-Safety/Japan-and-China-sign-food-safety-pact. 1 June 2010.
15. Shears P (2010) Food fraud – a current issue but an old problem. Br Food J 112(2):198–213
16. Gossner CM-E et al (2009) The melamine incident: implications for international food and feed safety. Environ Health Perspect 117(12):1803–1808
17. Grocery Manufacturers Association, AT Kearney (2010) Consumer product fraud: deterrence and detection
18. Greenhalgh M (2010) Chinese cooking oil found contaminated. Food Safety News, 22 Mar 2010
19. EU-bound sunflower oil 'deliberately' contaminated (2008) EUBusiness.com. Accessed 15 Feb 2010
20. Patton D (2008) 5 May EU, Ukraine take action against contaminated sunflower oil. FoodProductionDaily.com. Accessed 15 Feb 2010
21. Yan W (2010) Unsafe ingredient in some flours. China Daily. http://www.chinadaily.com.cn/china/2010-04/08/content_9699679.htm 8 Apr 2010
22. Layton L (2010) FDA pressured to combat rising 'food fraud'. Washington Post, 30 Mar 2010
23. Hennessey M, Kennedy S, Busta F (2010) Demeter's resilience: an international food defense exercise. J Food Prot 73(7):1353–1356
24. Nganje W, Richards T, Bravo J, Hu N, Kagan A, Acharya R, Edwards M (2009) Food safety and defense risks in U.S.-Mexico produce trade. Choices, 2nd Quarter 2009 24(2):1–8
25. Hartnett E, Paoli GM, Schaffner DW (2009) Modeling the public health system response to a terrorist event in the food supply. Risk Anal 29(11):1506–1520
26. Park J-S, Teren S, Tepp W, Beebe D, Johnson E, Abbott N (2006) Formation of oligopeptide-based polymeric membranes at interfaces between aqueous phases and thermotropic liquid crystals. Chem Mater 18(26):6147–6151
27. Hilgren J, Swanson K, Diez-Gonzalez F, Cords B (2007) Inactivation of *Bacillus anthracis* spores by liquid biocides in the presence of food residue. Appl Environ Microbiol 73(20):6370–6377
28. Venette S, Veil S, Sellnow T (2005) Essential communication resources for combating bioterrorism: some practical and generalizable recommendations. Commun Res Rep 22(1):29–37

Web Resources

http://www.fda.gov/Food/FoodDefense/default.htm
http://www.fsis.usda.gov/Food_Defense_&_Emergency_Response/index.asp
http://www.cis.utk.edu/safe/food_defense.shtml
http://wifss.ucdavis.edu/agroterrorism/index.php
http://www.ncfpd.umn.edu
http://www.who.int/foodsafety/publications/general/en/terrorist.pdf

Chapter 8
Establishing Database for Food Products and Ingredients to Strengthen Readiness in Food Terrorism Attack

Khalil I. Ereifej

Abstract Food tables such as food composition, food type and ingredients as part of food data base can provide detailed information which can be essential to nutritionists, food scientists, and many workers in a variety of fields such as epidemiological research, clinical practice and health promotion policy, food manufacturing, and to the State in fighting terrorism.

All citizens and dietetic professionals should be on alert in monitoring food safety from farm to fork. In addition to food products/ingredients data base tables, Food Nets and International cooperation, establishment a local Health Department to be the first to respond on any event might look like food terrorism.

Also, the individuals at the front line could observe food outbreak, increase in illnesses, unusual illnesses or irregular health behaviors. The dietitian in the front lines is at the key position to observe irregularity.

Quick communication and response will help to encounter the food product or ingredient, which is responsible or have been employed in a bioterrorism attack.

This presentation presents information on the History, Development, Present status of food data base for food products/ingredients and the recommended solutions in case of food bioterrorism threat.

Keywords Food data base • Food safety • Food terrorism

8.1 Introduction

Fresh and processed foods, along with food ingredients, provide nutrients and energy to consumers. They can be a target to terrorize unprotected civilians by contaminating them with potential biological agents such as bacteria, rickettsia,

---

K.I. Ereifej (✉)
Department of Nutrition and Food Technology, Faculty of Agriculture, University of Science and Technology, 3030, Irbid 22110, Jordan
e-mail: ereifej@just.edu.jo

M. Hefnawy (ed.), *Advances in Food Protection*, NATO Science for Peace and Security Series A: Chemistry and Biology 1, DOI 10.1007/978-94-007-1100-6_8,

Table 8.1 List of infectious and communicable diseases that are transmitted through handling the food supply and the methods by which such diseases are transmitted (Adapted from Khan et al. [1])

1. Pathogens often transmitted by food contaminated by infected persons who handle food
Calciviruses (Norwalk and Norwalk – like viruses)
Hepatitis A virus
Salmonella typhi
Shigella spp.
Staphylococcus aureus
Streptococcus pyogens
2. Pathogens occasionally transmitted by food contaminated by infected persons who handle food, but usually transmitted by contamination at the source or in food processing or by non-foodborne routes
Campylobacter jejuni
Cryptosporidium parvum
Entamoeba histolytica
Enterohemorrhagic *Escherichia coli*
Giadia Lambia
Nontyphoidal *Salmonella* spp.
Rotavirus
Taenia solium
Vibrio cholerae 01
Yersinia enterocolitica
3. Additional pathogens usually transmitted by contamination at the source, in food processing, or by non- foodborne routes
Bacillus cereus
Listeria monocytogenes
Vibrio parahaemolyticus
Vibrio vulnificus
Cyclospora cayatenensis

viruses or their toxins, or toxic chemical agents such as metals, cyanide, nerve agents or industrial chemicals and pesticides as shown in Table 8.1.

Food terrorism has been defined by the World Health Organization (WHO) as follows: "an act or threat of deliberate contamination of food for human consumption with chemical, biological, or radio-nuclear agents for the purpose of causing injury or death to civilian populations and/or disrupting social, economic, or political stability".

Food composition database provides detailed information on the levels of nutrients of the particular food or the ingredients incorporated with the food. Such information can be used by different systems such as epidemiological research, health promotion, clinical practice, food manufacturers, food policy makers and nutritionists.

The United States Department of Agriculture (USDA) and Agricultural Research Service developed and disseminated the widely used food composition database. The most popular tables are the Standard Reference (SR) and the Survey Database (SD). The two databases are alike in number of foods, but they differ in structure, and uses, as shown in Tables 8.2–8.5.

Table 8.2 Nutrients available and percent of foods with nutrient values in the USDA Standard Reference (SR) Database Version 13 (6,210 foods) and the 1999 Survey Database (~7,000 foods) (Adapted from Stumbo [2])

	Percent of foods with values	
Nutrient	SR	Survey
Macronutrients, energy, fiber and water (6 components)	100.00	100.00
Alcohol	0.55	100.00
Ash	99.61	N/A
Saturated fatty acids (sum of 13)	96.91	100.00
Monounsaturated fatty acids (sum of 5)	93.33	100.00
Polyunsaturated fatty acids (sum of 7)	93.33	100.00
Cholesterol	98.87	100.00
Phytosterol	9.76	N/A
Vitamin D	4.88	N/A
Vitamin A (μg RE) and (IU)	92.25	100.00
Carotene (RE)	N/A	100.00
Vitamin E (mg ATE)	55.73	100.00
Vitamin C	96.26	100.00
B Vitamin (B-1, 2,3,6,12 and folate)	92.65	100.00
Pantothenic acid	87.91	N/A
Minerals (Ca, P, Mg, Fe, Zn, Cu)	94.81	100.00
Electrolytes, sodium and potassium	97.75	100.00
Manganese	81.66	N/A
Selenium	80.47	100.00
Amino acids (18 individual)	68.32	N/A
Fatty acids (19 better- known compounds)	68.70	100.00
Fatty acids (6 less well known, i.e., margaric and behenic)	1.50	N/A

Table 8.3 Number of beef entries in the Standard Reference (SR) Release 13 and Survey Database Release 7 (Adapted from Stumbo [2])

Type of beef	SR	Survey
By use		
Roast	0	7
Steak	0	18
Stew	0	3
By cut		
Brisket, flank, shank	34	3
Chuck, round	215	0
Rib, round, loin, sirloin	300	6
General		
Not specified as to type	0	5
Composite	34	0
Ground	23	8
Sausage, luncheon meat	4	30
Variety (organ) meats	23	13
Commodity (USDA donated foods)	81	0
Total	718	92

Table 8.4 Number of hamburger entries in the Standard Reference (SR) Release 13 and Survey Database Release 7 (Adapted from Stumbo [2])

Restaurant	SR	Survey
Generic hamburger	10	38
Burger king	0	4
Hardees	0	1
Jack- in- box	0	10
Jimmy Dean	1	7
Mc Donald	0	4
Roy Rogers	0	2
Wendys	0	3
White Castle	0	2
All restaurants	1	33
Generic and restaurants	11	71

Table 8.5 Survey Database recipes (Adapted from Stumbo [2])

Cheese pizza, (NS[a] as to crust type)		Cookie, (NS[a] as to typ)
1.5 cup	Wheat flour	20 g shortbread cookie
1.5 tbsp.	Shortening	20 g vanilla sandwich
0.5 cup	Water	20 g sugar wafers
0.125 oz	Yeast	20 g animal crackers
0.5 tsp.	Sugar	20 g chocolate chip cookie
0.5 tsp.	Salt	Cereal, cooked, NS[1] as to type
4.0 oz	Mozzarella cheese	
0.5 cup	Canned tomatoes	
0.5 cup	Tomato puree	(loses 28 g water)
0.33 clove	garlic	0.5 cup cereal, oats
2.0 tbsp.	onion	1.0 tsp. salt
1.33 tsp.	olive oil	1.0 cup water
1.0 oz	parmesan cheese	
Potato chips		
Potato chips (single ingredient)		Fruit leather (include fruit roll- up, fruit snacks,
		Soda- licious, Farley's fruit snack, Gushers fruit snack, fruit by the foot, Betty Crocker fruit string thing)
		100 g fruit leather (single ingredient)

[a]NS = not specified

Stumbo [2] described the SR and the SD from a user perspective. The Standard Reference database has extensive details about different foods at the commodity level. The details given for beef are an example. In addition to beef composition, the different retail fat trim are stated (0, $^{1}/_{8}$ or ¼") and the grade (select, choice, prime) are listed, example data are presented in Table 8.3.

On the other hand, the Survey Database lists 25 steak entries and roast beef. The Standard Reference database usually does not suggest the suitable cooking method for beef. The Standard Reference database lists beef by cut (position on the carcass) and by grade and fat trim. These databases can be monitored and also the operational risk can be managed to increase the effectiveness of the operation against anticipated hazards and to minimize risks to acceptable levels. Management of operational risks associated with food commodities or their ingredients in SR or SD databases can provide more effective use of resources and can be used to improve food safety and security.

8.2 Food Control System

FAO/WHO defined food control system as "The Mandatory regulatory activity of enforcement by national and local authorities to provide consumer protection and ensure that all foods during production, handling, storage, processing, and distribution are safe, wholesome, and fit for human consumption, confirm to safety and quality requirements, and are honestly and accurately labeled as prescribed by law."

The existing food composition databases are needed to estimate intake of health promoting or hazardous component of foods [3]. European countries have their own food composition tables which were compiled using specific procedures developed by each country and traditions [4]. Those European food databases were not conceived to provide comparable data to other countries of the international community [5].

The food database in the United States, European countries, or other countries, show the food nutrients, stabilizers, gums, food additives, and preservatives, but they differ in language, classification, and description [6]. These differences make comparison on the international level difficult and imprecise as stated by EFCOSUM group [7].

Ireland et al. [4] concluded that foods can be made comparable at the raw ingredients level: therefore, software is quite essential to enable the conversion of foods as consumed to food at ingredients level. This software might also simplify the exchange of food database knowledge at the international level.

8.2.1 From Food Composition Tables to Food Composition Database

The sources of food chemical composition for many years were the food composition tables as shown in Table 8.6 which shows data on the chemical composition of some Jordanian cereals and legumes. Table 8.7 shows the concentration of some minerals in the same products.

Food composition data became of great size and spread due to employment of the computers and the internet. Also, food composition and the related literature became more accessible through the food database which has been accomplished via food composition tables [8].

Table 8.6 Energy and chemical composition of some Jordanian cereals and legumes as compared with Moroccan and FAO food tables [a] (Adapted from Ereifej and Haddad 2001)

Food type	Origin	Energy (KJ/100 g)	Protein (g/100 g)	Fat (g/100 g)	Carbohydrates (g/100 g)	Fiber (g/100 g)	Moisture (g/100 g)
Barley	Jordan	1,561 ± 1.7	14.1 ± 0.2	1.5 ± 0.3	75.7 ± 0.4	4.2.0 ± 0	4.5 ± 1.2
(*Hordeum vulgare*)	Morocco	1,274 ± 4.2	7.6 ± 0.1	2.2 ± 0.1	87.2 ± 0.1	3	7.8 ± 0.3
	FAO	1,565	11.1	1.8	84.6	–	10.1
Durham wheat	Jordan	1,515 ± 3.5	14.4 ± 0.1	1.4 ± 0.2	73.0 ± 0.6	3.3 ± 0.3	7.9 ± 0.5
(*Triticum durum*)	Morocco	1,682 ± 4.2	11.1 ± 0.2	2.1 ± 0.1	84.9 ± 0.3	–	8.3 ± 0.3
	FAO	1,615	14.9	1.9	80.8	–	15.0
Peas (*Pisum* spp.)	Jordan	1,544 ± 0.4	34.7 ± 0.3	1.0 ± 0.2	55.4 ± 0.8	2.3 ± 0.5	6.6 ± 0.2
	Morocco	1,665 ± 4.1	23.2 ± 0.4	1.9 ± 0.1	71.3 ± 0.2	–	8.3 ± 0.4
	FAO	1,594	24.7	1.2	69.7	–	11.3
Lentils	Jordan	1,565 ± 6.3	29.7 ± 0.3	1.5 ± 0.2	60.3 ± 1.6	5.0 ± 0.4	3.5 ± 0.1
(*Lens culinaris*)	Morocco	1,661 ± 4.1	22.0 ± 0.3	1.8 ± 0.1	72.3 ± 0.4	–	8.0 ± 0.3
	FAO	1,761	17.9	1.3	67.8	–	11.3
Chickpeas (*Cicer*	Jordan	1,682 ± 9.2	19.4 ± 0.8	6.9 ± 0.4	65.6 ± 0.3	3.3 ± 0.4	4.8 ± 0.2
arietinum)	Morocco	1,761 ± 8.3	17.9 ± 1	5.7 ± 0.3	73.1 ± 0.04	–	7.3 ± 0.5
	FAO	1,661	21.8	4.1	70.6	–	10.3

[a] Values are average of three replications and are mean ± SEM

Table 8.7 Minerals composition of some Jordanian cereals and legumes as compared with Moroccan and FAO food tables[a] (Adapted from Ereifej and Haddad 2001)

Food type	Origin	Ash (g/ 100 g)	Ca (mg/ 100 g)	P (mg/100 g)	Mg (mg/100 g)	K (mg/100 g)	Na (mg/ 100 g)
Barley	Jordan	2.4 ± 0.1	69.3 ± 3	179.0 ± 5	92.7 ± 2	612 ± 4	6.5 ± 0.8
(*Hordeum vulgare*)	Morocco	2.8 ± 0.1	69.9 ± 2.5	350 ± 6	135 ± 3	573 ± 11	20.3 ± 1.3
	FAO	2.6	67.7	279	–	–[b]	–
Durham wheat	Jordan	2.3 ± 0.1	29.5 ± 0.6	265.3 ± 3	85.1 ± 4	625 ± 8	7.9 ± 0.9
(*Triticum durum*)	Morocco	2.0 ± 0.2	40.1 ± 0.5	392 ± 6	137 ± 2	515 ±7	5.6 ± 1.2
	FAO	2.4	55.8	349	–	–	–
Peas (*Pisum* spp.)	Jordan	3.8 ± 0.3	71.7 ± 3.5	771.7 ± 5	184.5 ± 2	1,050 ±22	54.9 ± 6
	Morocco	3.0 ± 0.1	106 ± 3	393 ± 7	143 ± 3	1,111 ± 26	2.9 ± 0.3
	FAO	4.0	101	429	–	–	–
Lentils (*Lens culinaris*)	Jordan	3.5 ± 0.2	42.3 ± 4	458.5 ± 10	12.9 ± 2	38.1 ± 3	78.6 ± 5
	Morocco	3.2 ± 0.3	97.9 ± 4.8	315 ± 331	119 ± 1	548 ± 21	30.4 ± 10.6
	FAO	3.2	71.0		–	–	–
Chickpeas	Jordan	3.1 ± 0.1	225 ± 0.1	2.2 ± 0.2	1.4 ± 0	13.6 ± 1	264 ± 8
(*Cicer arietinum*)	Morocco	2.7 ± 0.2	182 ± 3	344 ± 6	151 ± 3	1,119 ± 20	8.01 ± 1.3
	FAO	3.6	280 ± 2	301	–	–	–

[a]Values are average of three replications and are mean ± SEM
[b] – Not available

Food databases are well established in Europe [9, 10]. The early American tables perhaps were these published in the United States in 1896 [11]. Germany started publishing the food composition tables in 1878 [12]. Other nations followed at a later time such as the Netherlands [13].

At present time, international work started in 1984 to harmonize food composition data with efforts to include food components identification and interchanging data by establishment of the international network of food data systems (INFOODS) under the auspices of the United Nations University [14]. Further work on data harmonization and its application in Europe to benefit the whole World utilizing food composition database is undertaken by EuroFIR (European Food Information Resources). EuroFIR is a network supported financially by the European Commission's "Food Quality and Safety Priority" of the Sixth framework program for Research and Technological Development [8].

8.2.2 Development of Food Composition Database

There are a lot of details reported in the literature in regard of the chronological development of food composition tables and their moving away to food composition database.

In this paper we would like to start from the first milestone and to cover the key events in the development of food composition database:

- In 1818: The first Food Composition table in France was published, data were generated from investigation of food supply in prison.
- 1878: The first European Food composition table as known today was published in Germany.
- 1896: The United States published the earliest table. 2,600 analysis of foods were incorporated (meats, cereals, fruits and vegetables, and some processed foods). Values were presented for water, protein, fats, and carbohydrates (by difference) for each food commodity. Later, crude fiber values were added (Tables 8.6 and 8.7 are examples).
- 1921: FAO Published the first food composition table for 900 foods which was accomplished by a chemist (Captain Plimmer) after the First World War during 1914–1918 [15].
- 1949: FAO published "Food Composition Tables for International Use". They were published to help and to assist in assessment of food availability at the global level. Calculation of energy, protein, fat on a per capita basis, using food balance sheet appeared in these tables [16].
- 1949–1950: FAO has been successful in producing and disseminating food composition tables and related information at the International level, and published food composition tables during this period.
- 1960–1970: FAO published regional food composition tables for Asia, Latin America, and Africa.

- 1980: United Nations University (UNU) responded to the continued interest in food composition tables by setting the International Network of Food Data Systems (INFOODS).
- 1982: FAO published food composition tables for the Middle East. Most of the work was completed in 1970s, and food composition table remained among FAO bestsellers, then FAO reduced the work on food composition tables.
- In 2001 Chemical Composition of Selected Jordanian Cereals and Legumes as Compared With the FAO, Moroccan, East Asian, and Latin American Tables for use in the Middle East was published [17].
- In 2010: NATO held an advanced research workshop on "advances in food security and safety against terrorism threats and natural disasters" in Cairo, April 13–15 on the matter.

8.2.3 Food Composition Data

As Lupien [18] stated that the complexity and nature of food composition data around the World requires genuine collaboration between all institutions and sectors generating food composition data. Building a new nutrient database is becoming difficult, especially in the developing countries, and that is due to lack of financing the detailed food analysis.

Therefore collaboration work gives the opportunity to reduce the cost, and food composition data can be provided to all developing countries through the Food and Agricultural Organization (FAO).

Food composition data which are provided to policy makers should be very accurate to avoid creating interventions and to ensure that the data will not affect the policy decisions.

We agree with FAO [19] and Lupien [18] that the basic objective is to generate (accurate food analysis), disseminate, use and promote highly accurate and precise food composition data by those who are researchers in the field, policy makers, and practitioners in public and private sectors around the World.

The coordination role among Eastern Europe, Anglophone, Francophone Africa, Asia and Latin America can be played by FAO to coordinate the work on food composition data, database, food type, and food ingredients or in case of food terrorist attack when a country is hit. FAO can fulfill the coordination role due to the following reasons as stated by Lupien [18]:

1. FAO can coordinate the international food activities-due to its United Nations' mandate such as food standardization (Codex Alimentarius), food safety, food quality, and international food trade.
2. FAO also has the broad international mandate which can allow it to involve in all food related development issues and food bioterrorism should be one of them.
3. Communication system has been established between FAO and governments to promote regional and international cooperation.

4. FAO circulates its publication as mentioned previously; FAO circulated its own food composition data.
5. FAO also is authorized to take actions on interdisplinary issues which might some times require an open forum to formulate solutions.

Database for food type and food ingredients is necessary and could be built to encounter food related problems such as safety when using food as a vehicle for massive death. A strategy such as a dynamic system to handle food composition data information or other issues such as food bioterrorism can perform some important functions such as follows:

1. As a result of changes in food product formulation, food processing technologies, food varieties and types, food production systems, developments of analytical techniques and food safety problems, there should be a continuous generation, distribution, and revision to all these issues.
2. Standards and procedures to specify the minimum quality criteria required for food and ingredients should be updated continuously.
3. A structure of governmental committees and organizational representative should be maintained to approve standards, specification procedures, and priorities for food composition, food types, ingredients, and procedures to encounter any food bioterrorism attack.

In 2001, the Acting Principal Deputy Commissioner of the FAO, Bernard A. Schwetz emphasized the importance of cleansing raw food products, especially in the context of food bioterrorism. It should be also recognized the importance of strengthening the food safety systems, due to continuous change of our World; we need to change with it and to build a strong and credible food safety systems to address the full range of food safety issues.

The United States has introduced two systems and are vital in encountering the agents of food bioterrorism weapons such as anthrax, botulism, plague, smallpox, and tularemia or those listed in Table 8.1. Those two systems are:

1. FoodNet: the Foodborne Disease Active Surveillance Network. It is part of CDC's Emerging Infections Program which is a project of the CDC, USDA, FDA, and nine states. This system tracks the incidence and trend in foodborne illness and systematically conducts investigations in order to assist health officials in understanding the epidemiology of foodborne disease in the United States. This system can benefit the developing countries if it is employed outside the USA.
2. PulseNet: A system developed by the CDC to enable the national public health laboratories networks to "fingerprint" bacteria that might be foodborne and compare results with database. This is a collaborative work between CDC, FDA, and USDA.
3. This system – PulseNet – permits early detection of foodborne illness outbreaks which have not been detected in the past.

8.2.4 Additional Database

Ottley [20] discussed the compatibility issues associated with food composition tables. Some problems might arise especially when food data are needed to be comparable in more than one country, where researchers or legislators need that. Ottley [20] concluded that such incompatibility confounds international researchers and reduces the merit of scientific research findings of an association between intake of a food component or food ingredient and a health outcome.

International prospective food investigators should realize the following differences among food composition tables such as type of food and nutrient covered, food description, sampling method, and analytical technique applied to generate the data, data of quality assessment, method of nutrient level determination, calculation, and finally unit used to express the concentration or level of food component.

We also would like to state that people around the world differ in their food habits, food, or meal preference and also the amount of food consumed by the individual. Some foods such as bread or rice are staple foods for the majority of the people on the Globe. Therefore, these foods, their products or the ingredients which are used in their products might be a target by terrorist groups for massive killing.

The FAO might encourage all nations to produce their own food products database along with food ingredients database and/or adapt the food database systems and tools which are employed in Europe or the United States.

8.2.5 Food Product Database

Food bioterrorism is the threatened use of microorganisms or biologic toxins to kill or incapacitate people, animals, or corpses. The advantages of food bioterrorism are easy to obtain and inexpensive to produce. Such weapons can create panic, overwhelms the media and the perpetrator can escape easily. Therefore, the country's emergency response and the public health infrastructure should be adequate and ready to act when bioterrorism event occur.

The early attempts of biological and chemical terrorism probably started in 429 BC when Spartans ignited pitch and sulfur to create toxic fumes during the Peloponnesian War. In 1345, Mogols catapulted corps contaminated with plague over the walls into Kaffa.

During the French-Indian War in 1763, the British gave smallpox contaminated blankets to hostile the Indian tribes.

During the First World War (1914–1918) the biological and chemical weapons were widespread. But in June 17, 1925, Geneva Protocol prohibited the use of asphyxiating, poisonous, or other gases, also, the bacteriological methods in warfare.

In 1932 Japan started biological war program research, and in 1940 Japan waged biological war on Manchuria. The United States responded by starting an offensive biological war program in 1942.

During 1969–1970, President Richard Nixon ordered unilaterally dismantling the US offensive biological and chemical weapons. Since 1972 to present, there were lots of suspected or proven violations of biological and chemical weapon events around the World.

8.3 Incidents of Food Bioterrorism

8.3.1 Bioterrorism Agents

The food bioterrorism agent can be grouped into categories. Category A agents: which have the following agents: (require special action): Anthrax (*Bacillus anthracis*), Botulism (*Clostridium botulinum toxin*). Plague (*Yersinia pestis*), Smallpox (*Variola major*), Tularemia (*Francisella tularensis*), and Viral hemorrhagic fever (e.g., Ebola, Marburg, Lassa, Junin).

Category B agents have the following agents: (require specific enhancement): Brucellosis (*Brucella*), Epsilon toxin of *Clostridium perfringens*. Food threats such as: *Salmonella*, *E. coli:* 0157: H7, *Shigella*, *Vibrio cholerae*, *Cryptosporidium parvum,* Glanders (*Burkholderia mallei*), Melioidosis *(B. pseudomallei*), Psittacosis (*C. psittaci*), and Q Fever (*Coxiella burnetii*). Ricin toxin can be obtained from Castor beans. *Staphylococcus* enterotoxin B. Typhus fever (*R. prowazekii*), Viral encephalitis (such as: EEE), Water threats VEE, WEE.

Category C agents which might be considered the future threat: Nipah virus, Hantavirus, Tickborne HFVs, Ttickborne EVs, Yellow fever and Multidrug resistant organisms.

Individuals, groups, or governments might think food biological terrorism is an unlikely threat. Those who believe in that have their own reasons such as the use of biological-chemical weapons is infrequent, there are technical difficulties in obtaining or using such weapons also; such arms are hazardous to perpetrators. Others believe that food bioterrorism is a real threat, in spite of the low probability but high consequence threat. Food bioterrorism happened somewhere some time, and it will happen again. Therefore, preparedness and response planning against food bioterrorism are no longer a national priority; it should continue following an all-hazard philosophy. During 2002–2008, CDC issued the emergency supplemental public health preparedness and response to bioterrorism which comprise the following:

1. Preparedness Planning and Readiness Assessment.
2. Surveillance and Epidemiology Capacity.
3. Laboratory Capacity-for Chemical Agents.
4. Laboratory Capacity-for Biological Agents.
5. Health Alert Network.
6. Communication and Health info Dissemination.
7. Education and Training.

8.3.2 Fighting Food Bioterrorism

The national FDA is the main food regulatory agency, which should be authorized to register food plants, food products, and food ingredients. The regulatory food agency should be the only authorized governmental body to issue a prior notice to import foods or food ingredient. Also, the regulatory agency should be given the authority to keep records for food products, ingredients and to have an access to all records always or under certain circumstances.

Food and ingredients records are of major importance; therefore persons who handle foods and ingredients must maintain records of the immediate, previous source and immediate subsequent recipients of the food or ingredients.

Food ingredients manufacturer, food processor, food packer, transporters, distributers, receivers, holders, and importers also must keep records.

If a food system is vulnerable to sabotage, the points of vulnerability are its nodes or links.

What could we do to interrupt food bioterrorism attack? There are many steps can be taken on the governmental or public levels, national, and international levels. We would like to point out some approaches which can be done to prevent or encounter food bioterrorism.

8.3.3 On the International Levels

Countries such as the under-developing countries can adopt the American Anti Bioterrorism Acts which was issued in 2002. Adoption of new food regulations can be very useful to control and prevent food frauds or anybody to reach food products and ingredients for sabotage purposes.

One of the major important steps which can be taken on the international level is to adopt an international trade agreement to control import-export of all food products and food ingredients. Such international trade agreement can authorize the local food agencies to issue previous permission for food import from authentic sources.

Food and drug agencies in each country should be established and should be authorized to inspect the food plants, products, and ingredients and the related records at any time/period.

Because food testing is expensive but it is essential during food terrorism attack or during surveillance, the necessary funds for food testing to find out the bioterrorism agent which was used by the bioterrorist group should be drawn from governmental sources or funding agencies.

International cooperation between world wide organization, countries, food societies, manufacturers, exporters, importers, and international food regulatory agencies should be brought together into full cooperation.

8.3.4 Surveillance

Many private enterprises can develop several approaches for food safety and food traceability; food product and food ingredients safety are very important: therefore, food safety systems must be adapted and practiced in food plants and through food supply chain which include farmers, distribution centers, food processor, packers, shippers, exporters, importers, food service systems such as retailers and restaurants.

The ISO 22000 and 22005 which include all the food safety systems and cover up the standards for food safety and traceability, the safety approaches should cover up the recall procedures for food products, ingredients, and contaminated food during terrorist threat. Also should cover the food consultant systems such as food safety inspection, food safety compliance, body personnel, and any food products or any institutions wish to choose food traceability and surveillance systems. Those might include farmers, specialists in food safety, food distribution, and food relating outlets.

In case of bioterrorism attack, surveillance might involve important steps such as establishing a local health department. This department can be the first to respond on any event related to food terrorism. The individual at the frontline can respond about his/her observation of food outbreak, unusual increase in illnesses or abnormality in illness, or any irregular health behaviors.

International cooperation is very essential to adopt similar systems to encounter food bioterrorism. The quick responses for the front lines help for quick recall, testing or taking immediate action to fight the food bioterrorism threat.

8.3.5 Traceability

It is obvious that processed food is different than fresh food; the raw material which is used to produce processed products might come from different sources, and that is true also for all the ingredients which are added to the food product. Raw materials and ingredients undergo changing during manufacturing operations such as mixing and other food processing steps.

The raw material, the product itself, and the ingredient or ingredients might be subjected to food threat; therefore, to counterpart the food terrorism threat should rely on a reliable traceability procedure. It is very difficult to utilize a traceability procedure for each raw material or each food ingredient which comprise different products. A code bar has been suggested to be employed and declared on the food label; this might cause a trouble or difficulty because one finished product might contain several raw materials and several food ingredients, there must be a code bar for each one to trace the agent in the unlikely event of food terrorism threat to encounter the threat itself.

To make our food safe and protected on the national or the international levels, we should implement all food safety systems (food quality assurance, food quality control, good manufacturing practices, HACCP, ISO 22005, food security, and food defense). In addition to all of that, there must be traceability and a good and swift recall procedure.

Unfortunately, we have succeeded in all implementing all of these food safety systems, but food traceability is still in progress. Food traceability requirements are already established in Europe and America. The Asian countries are moving fast in establishing food traceability requirements. Other countries can adopt similar systems to protect their food from bioterrorism threats.

We would like to emphasize that food safety and traceability systems can involve the revision of food and drug agencies directions, bar code and annual systems, the international standards for food safety and traceability.

Finally, the adoption of food safety systems from Asia or other countries advanced in food inspections, food auditing, and laboratory testing, should be directed to work toward food safety, surveillance, food protection and preventing terrorist threat.

8.3.6 *Some Useful Tips for People During Bioterrorism Attack*

In the unlikely events of food terrorism attack, people can help themselves, the community, and the state by upraising the following tips and that will help the people in the front lines such as the Public Health Department, dilatations, and food service systems people.

People can be trained to enhance their ability to identify food safety risks through media, workshops, and lecturers to make them familiar with the concepts of food safety and the risks involved.

Inspection systems for food plants, warehouses, and transportations for food type and food ingredients can be improved and make everybody responsible for food protection from farm to fork.

State laboratories, private sectors and laboratories based in food plants can adopt and strengthen quick and accurate methods to measure food safety activities. Food quality assurance, good manufacturing practice, hazard analysis of critical control points, and ISO 22005 can be adopted by food plants for food products and ingredients to insure continuous food safety and protection.

Consumers are very helpful in preventing food terrorism specially when they are trained or given the knowledge for basic food safety behaviors and that will help to reduce and minimize the risk of foodborne diseases. A standard food safety guidelines or day-to-day bases can be very useful.

We would like to encourage consumers only to acquire and purchase food from reputable vendors which are usually controlled by regulatory agencies. Also, to encourage consumers to check food for its intact packaging and cans to be washed before opening, and keeping away debris from can lid from falling into food.

In the unlikely event of food threat or attack, people are encouraged to keep a few days worth of food, water, and medicine because distribution of these items might take some times after bioterrorism attack.

People are encouraged to utilize all collective resources on foodborne diseases, bioterrorism, and food safety such as FoodNet. The Foodborne Disease Active Surveillance Network and PulseNet, which was developed by CDC, because bioterrorism attack is silent and the effects might take several days to be noticeable. People are encouraged to convey any useful information to the Department of Homeland Security, Health Department, Food Agency, or any bureau in charge.

We would like to suggest some solutions in case of food terrorist threat or a catastrophe; some of these suggested solutions are listed below:

1. Enhancing the ability to identify food safety risks.
2. Improving the inspection of domestic and imported foods.
3. Strengthen science-based methods that measure the out come of food safety activities.
4. Adhering to basic food safety behaviors will help reduce the risk of foodborne disease.
5. Additional standard food safety guidelines on day-to-day basis.
6. Accept only food from reputable vendors to take advantage of public controls from regulatory agencies.
7. Check for intact packaging.
8. Wash cans before opening.
9. Keep 3 days or more worth of food, water (one gallon of water per person per day) and prescription medicine available. After bioterrorism emergency, the distribution of these items might take some times to coordinate and organize.
10. Interact and consult with FoodNet and PulseNet.
11. Bioterrorism attack would be silent, and the effects might not be visible for several days.

Acknowledgments The author would like to thank the NATO Science for Peace & Security Program for supporting the Advanced Research Workshop in Food Security and Safety Against Terrorist Threats and Natural Disaster, National Research Centre in Cairo, Egypt, also The Agricultural Technology International for their support in holding the workshop in April 13–15, 2010, in Cairo, Egypt. The efforts of all other scientists participated in the workshop and those who helped to make it successful are acknowledged.

References

1. Khan AS, Swerdlow DL, Juranek DD (2001) Precautions against biological and chemical terrorism directed at food and water supplies. Public Health Rep 116:3–14
2. Stumbo PJ (2001) Structure and uses of USDA food composition databases. J Food Comp Anal 14:323–328
3. Greenfield H, Southgate DAT (1992) Food composition data: production, management and use. Elsevier, London
4. Ireland J, van Erp-Baart AMJ, Charrondière UR, MØller A, Smithers G, Trichoopoulou (2002) Selection of a food classification system and a food composition database for future food consumption surveys. Eur J Clin Nutr 56(suppl 2):533–545

5. Deharveng G, Charrondièr U, Slimani N, Southgate DAT, Riboli E (1999) Comparisons of nutrients in the food composition tables available in the nine European countries participating EPIC. Eur J Clin Nutr 53:60–79
6. Slimani N, Ferrari P, Ocke M, Welch A, Boeing H, Van Liere M, Pala V, Amiano P, Lagiou A, Mattisson I, Stripp C, Engest D, Charron-diere R, Buzzard M, Van Stevenren W, Riboli E (2000) Standardization of the 24-hour diet recall calibration method used in the European Prospective International into Cancer and Nutrition (EPIC): general concept and preliminary results. Eur J Clin Nutr 54:900–917
7. EFCOSUM group (2001) European food consumption survey method. Final report. TNO report V3766. TNO Nutrition and Food Research, Zeist
8. Burgos MA Martinez, Martinez-Victoria I, Mala R, Farran A, Farre R, Ros G, Yago MD, Audi N, Santana C, Lopez Millan MB, Romos Lopez S, Manas M, Martinez-Victoria E (2009) Building a unified Spanish food database according to EuroFIR specifications. Food Chem 113:784–788
9. Paul AA, Southgate DAT (1978) McCance R. A. and Widdowson's The composition of foods, 3rd edn. HMSO, London
10. Church SM (2006) The history of food composition databases. Nutr Bull 31:15–20, British Nutrition Foundation
11. Atwater WO, Woods CD (1896) The chemical composition of American food materials. US Office of Experiment Station. Experiment Stations Bulletin 28. Government Printing Office, Washington, D.C. Available at http://www.nal.usda.gov/fnic/foodcomp/data/classics/index.html
12. Konig J (1878) Chemie der Menschlichen Nahrungs-und Genussmittel. Springer, Berlin
13. van Eekelen M, Jassen BCP, Straub J (1941) Voedingsmiddelentable
14. Scrimshow N (1997) INFOODS: the international network of food data system. Am J Clin Nutr 66(suppl):S1190–S1193
15. Plimmer RHA (1921) Analyses and energy values of foods. HMSO, London
16. Southgate DAT (1993) The composition of foods. In: Ashwell M (ed) McCance and Widdowoson: a scientific partnership of 60 years. British Nutrition Foundation, London, pp 69–77
17. Ereifej KI, Haddad SG (2001) Chemical composition of selected Jordanian cereals and legumes as compared with the FAO, Moroccan, East Asian and Latin American tables for use in the Middle East. Trends Food Sci Technol 11(9–10):374–378
18. Lupin JR (1997) The Food and Agriculture Organization food-composition initiatives. Am J Clin Nutr 65(suppl):1194S–1197S
19. FAO, Food and Agriculture Organization/US Department of Agriculture (1982) Food composition tables for use in near east. FAO, Rome
20. Ottley C (2005) Food composition database. Trends Food Sci Technol 16:507–509

Chapter 9
An Introduction to Traceability for Processed Foods*

Thomas A. Butterworth

Abstract Traceability systems that are modelled after fresh produce are inadequate for processed foods. It is a common mistake since regulations and current scientific and trade literature on traceability generally do not specify sector-specific methods of implementing traceability. Most processed foods require more complex operations than fresh produce, and implementing traceability along with other quality initiatives will necessarily be more complex. Most processors recognize the increased complexity caused by the use of multiple ingredients for a single product and the manufacture of several products from different combinations of ingredients. As such, most processors believe they have adequate traceability systems in place. However, keeping track of the inputs and outputs are only parts of an effective traceability system. Processors tend to assume that internal traceability is well-known and controlled since yield, costs and certain quality parameters are well-known and controlled. However, traceability systems sometimes require unique information from processors since in-process mixing blurs the straight line that usually exists between inputs and outputs for fresh produce. Three different types of in-process blending can occur: blending that is "purposeful," (ex. the manufacture of olive oil), blending that is "continuous," (ex. a continuous evaporator) and blending that is "idiosyncratic" (ex. that which occurs in an aseptic surge tank). This third type of blending defies analysis for traceability. Each step in a process must be examined to determine if blending takes place, what type of blending it is, and how to accommodate the traceability system to it.

Keywords Food safety • Traceability • Processed foods

---

*Much of this information was developed by the author at the Egyptian Traceability Centre (ETRACE), 106 Gamet El Dewal El Arabia Street, 5th Floor, Mohandessin, Giza EGYPT; www.etrace-eg.org ETRACE was supported by UNIDO.

T.A. Butterworth (✉)
Austin Food Tech Inc., 9 Yemen Street #30, Mohandessin, Giza 12411, Egypt
e-mail: foodtech@alum.mit.edu

M. Hefnawy (ed.), *Advances in Food Protection*, NATO Science for Peace and Security Series A: Chemistry and Biology 1, DOI 10.1007/978-94-007-1100-6_9,

9.1 Introduction

Most processed foods require more complex operations than fresh produce, and implementing traceability along with other quality initiatives will necessarily be more complex. One way of looking at this increased complexity is to consider the apple. A fresh-market apple may go through many processes before it appears in the market. However, it is still an apple, not fundamentally different from when it was on the tree. It represents the ultimate in batch systems. A good traceability system could determine which tree it came from and when. However, if an apple goes into a process for the manufacture of a shelf-stable cocktail juice, that apple has almost disappeared. The apple has taken on a new form. It is a small and invisible component of a product with many ingredients, and that product most likely contains juice from apples from different trees, farms, and even regions and countries.

The application of traceability has gained some degree of acceptance among food **processors**, in part because of EU and US regulations. As such, most **processors** believe they have adequate traceability systems in place. Many do, but many have systems that are modelled after fresh produce and are inadequate for processed foods. It is a common mistake since regulations and current scientific and trade literature on traceability generally do not specify sector-specific methods of implementing traceability.

Traceability is part of bigger systems, such as **ISO** 9001, or **HACCP**, and more recently, is also a mandate in **ISO** 22000 and described in **ISO** 22005. Any manufacturer claiming compliance with such systems must have traceability in place.

9.2 What Is Traceability?

Traceability is a simple concept, one that most people are likely to understand to some degree. According to European Council regulation (EC) 178/220,[1] "traceability" means:

> …the ability to trace and follow a food, feed, food-producing animal, or substance intended to be or expected to be incorporated into a food or feed through all stages of production, processing, and distribution

In commercial practice, traceability includes details of what has happened to the food (its processing history) as well as the source of the raw materials and the recipient of the finished product.

Traceability has two components:

- Tracking: The ability to follow the path of a specified unit and/or **lot** of trade items downstream through the supply chain as it moves between trading partners. Units are tracked routinely for availability, inventory management, and logistical purposes.

---

[1] http://eur-lex.europa.eu/LexUriServ/LexUriServ.do?uri=CELEX:32002R0178:EN:HTML

- Tracing: The ability to identify the origin of a particular unit located within the supply chain by reference to records held upstream in the supply chain. Units are traced for purposes such as recall and complaints.[2]

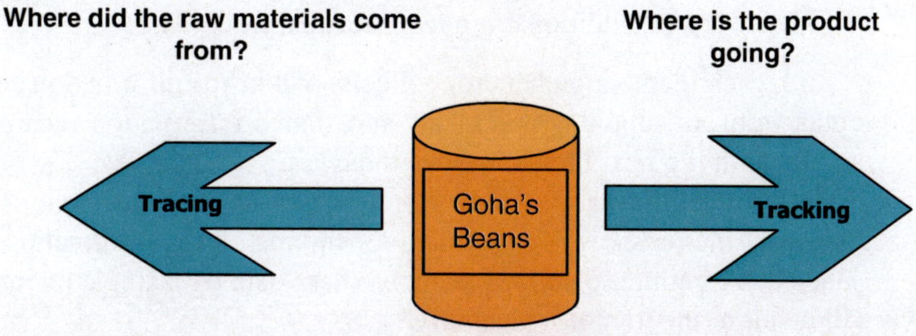

Internal Traceability

Traceability systems should be able to provide the following:

1. The identity of units or batches of all raw materials, ingredients, and primary packaging materials supplied
2. **Information on how, when, and where they were incorporated into finished** products[3]
3. The identity of units or batches of all products manufactured and to whom these are supplied
4. A documentation and record keeping system linking these data

Tracking and tracing a processed food requires that the finished product bear a unique **lot identifier (production code** & **label)** to define (after examination of the **ultimate processor's** records) the first three characteristics, above. This unique identifier does not change once the product is packed into individual coded containers. As the product moves down the value chain from the **ultimate processor,** it is only the **production code** and **label** that begin to identify the elements of the traceability system.

In a **one up/one down** model of traceability, no single member of the supply chain holds all the information. However, the onus is on the **ultimate processor** since:

- It is usually the one member of the value chain that brings most of the inputs together and has the most issues with **internal traceability**.
- When the product leaves the premises of the **ultimate processor**, the traceability game is essentially over, since
  - The **production code** and **label** should contain all the information needed to start to determine inputs and how they were processed,

---

[2] Canadian Food Traceability Data Standard, version 2.0 (2006); Note: GS1 Canada is the industry custodian of the Can-Trace standard.

http://www.can-trace.org/portals/0/docs/CFTDS%20version%202.0%20FINAL.pdf

[3] Since traceability systems should operate in "real time," they should be able to reveal the history of any given component in a process at any point in time. Depending on the extent of the traceability system, it can not only reveal which lots of product were manufactured from which lots of raw materials, but it can also reveal the source and causes of out-of-specification product.

  – The value chain members downstream from the **ultimate processor** will have no need for any product details except what is on the **production code** and the **label** and,
  – Once the **ultimate processor** packs the product into an individual container, there are no additional inputs or further processing.[4]

Still, each member gathers (or collects) and keeps information about production inputs from its suppliers and keeps and shares information regarding production outputs with the next member down the chain.

Collecting information from suppliers and sharing information with the next trading partner ensures a mechanism for linking all the information throughout the value chain. Failure to collect, keep, or share data by a single member of the chain will result in the loss of traceability.[5]

Traceability systems are thought to have three dimensions: **breadth**, **depth,** and **precision**. **Breadth** is the amount of information the traceability system demands. The **depth** of a traceability system is how far backward or forward the system traces or tracks (such as **one up/one down**). Although the **one up/one down** model complies with European Council regulation (EC) 178/220, individual customers may demand more. **Precision** is the degree of assurance that a traceability system can pinpoint the movements of a particular **ingredient** to a single **lot** of finished product. A traceability system with very high **precision** could likely trace an apple, in the trade, to its tree. A less precise system for apple juice may trace apples only to one or more of several growing areas. Conversely, in a less precise system, a single **production code** of apples may be tracked to one (or more) of several **production codes** of finished product. Reduced **precision** in traceability results in:

- Single **lot** numbers of raw materials ending up in more than a single **production code** of finished product (This may be thought of as dilution of a single **production code** of raw materials).
- More finished product codes containing more **lot** numbers of the same **ingredient**
- Uncertainty about the extent of the dilution of individual **lot** numbers of raw materials into more than a single **production code** of finished product.
  – This uncertainty leads to uncertainty about whether or not a small quantity of a given **lot** number of raw material is contained in a given **lot** number of finished product.[6]

---

[4] The exceptions are products manufactured for industrial uses by **intermediate processors**, such as tomato paste in bulk.

[5] For a discussion on how to handle dealers, see "Dealers and Other Middle-Men," below.

[6] We will see later that traceability systems for most processed foods are necessarily less precise than those for fresh produce. Traceability procedures are not changed by this limitation. The implications of the uncertainty are felt mostly during a recall where lots known to contain problems are removed from the trade. Uncertainty forces the processor to recall products not known, but suspected, to have problems.

The characteristics of a traceability system may be driven by characteristics of the raw materials and products. For example, raw materials and products that are more susceptible to pathogens, prone to chemical contamination, or intended for at-risk consumers such as children may have traceability systems that are broader, deeper and/or more precise. Also, a company that makes several products may wish to give priority in traceability to those products with the greatest sales volume.

Ultimately, the complexity and even the very existence of a traceability system will depend on a cost/benefit analysis on the basis of real and perceived advantages, outlined below in "Benefits."

9.3 What Are the Benefits of Traceability?

Traceability is a tool to help achieve broader management objectives, but management must define these objectives. What follows are some examples of such objectives:

The most important objective for traceability might be to meet quality, customer, regulatory and industry standards. Failure to meet these standards can result in lower prices for a **processor's** products or even the loss of some business altogether.

The second most important objective might be to limit the scope, and therefore the cost, of a product recall by determining the origins of finished products. It can also help trace and assign responsibility for defective products or raw materials.

Traceability can have other roles to fulfill depending on the company's objectives for a traceability system and its willingness to invest in it:

- To help support product claims and provide information to customers, for example to authenticate organic, non-GMO or country-of-origin claims
- To identify and implement corrective actions to regain control of a problematic process
- To provide information for the disposition of any non-conforming product, including recall if necessary.
- To serve as a tool for process control in areas such as inventory control and quality control.
- To provide a mechanism for providing product information quickly, to regulators or customers.

Traceability assures that regulatory authorities have the ability to trace all the steps taken during the preparation and distribution of food products as well as the sources of all the raw agricultural commodities, **ingredients,** and primary packaging materials. This ability is particularly useful in cases that might present health emergencies and resulting removal of certain product from trade. Traceability, therefore, assists risk management systems in that it ensures transparency, improves supply-chain management, and allows smooth and cost-effective exchanges of information.

The extent of a traceability system will be a balance between costs and benefits accrued. Any traceability system and each component of that system must have a measurable economic benefit. Otherwise, it will serve no purpose.

It is sometimes tempting to suggest that traceability assures food safety. Traceability can be a tool for food safety but only that. If Quality Control and **HACCP** systems are not adequately implemented and linked to inputs and outputs, a traceability system becomes little more than an after-the-fact trail of paper useful only in assigning blame if a problem arises. However, if hazards[7] are monitored, recorded, controlled, and LINKED to inputs and finished products, then the traceability system can become a powerful tool for food safety.

9.4 Internal Traceability

Internal traceability is the ability to track what happens to raw materials, **ingredients**, primary packages, and finished products inside the **processor's** operation. **Internal traceability** matches inputs to outputs. This means that when 1,000 kg raw material (or 10,000 primary packages) goes into a process, the records must be able to account for[8] the disposition of all this material. In the case of processed foods, this "accounting for" can be especially difficult since they involve operations which seemed designed to obscure their traceability. The biggest difference between traceability for fresh produce and processed foods is that **internal traceability** is more complex for processed foods:

- Processed products usually have more than one **ingredient**
- The processes tend to be more complex, and
- During manufacture, **ingredients** from different **lot** numbers are frequently blended into finished products with the same code.

Additional complexity from more **ingredients** and more complex processes should be self-explanatory. What may not be so obvious is the source and the necessity of blended **lot** numbers.

9.4.1 Blending of Product Codes

Most food processing operations result in the blending of different **production codes** of the same agricultural commodities and **ingredients**. This blending is what makes the **internal traceability** of processed foods more difficult.

[7] Common major hazards include aflatoxin, pesticide residues, heavy metals or pathogens. Other possible hazards and concerns that could be linked to the traceability system include allergens and GMO inputs.

[8] This not only means in the accounting sense, but also accounting for which materials ended up in what production codes.

Some of this blending results from the nature of a **continuous** process where there is no stoppage between different **production codes** of raw materials. For example, if an olive oil process is receiving and processing olives continuously and the **lot** number of the incoming olives changes, the process is not stopped and restarted for the new **lot**. Some mixing of different **lots** of olives may occur, especially during crushing and centrifugation. The end result is that some fraction of the finished product will contain two **lot** numbers of incoming fruit. Another example of **continuous** blending is as simple as a casing operation. After filling the last whole case in a **production code**, there will usually be a few primary containers of the same **lot** code that can not fill a case. These containers are not thrown away, and a case is not partially filled. The case is filled to its capacity with these "left over" containers as well as some containers from the next **production code**. This case will contain two **production codes**. The same analysis may be applied to palletizing. **Continuous** mixing usually results in the blending of two or three **lot** codes only. However, **continuous** mixing could affect, for example, all the production of a continuous evaporator from the time of a problem to the next **CIP**.

Another example of **continuous** blending that may not be so obvious is a size grading operation. Okra is graded into five sizes, 0–4. The middle grades, 1, 2, and 3 may predominate, and a process-able quantity (usually considered to be about 1,000 kg) usually will be produced from a single **lot** number of okra. However, the extremes of size (0 and 4) may require that several **lot** numbers be graded before a quantity large enough to be processed can result. Many **processors** ignore this problem, consequently, traceability at this point is lost.

Blending may also be **purposeful**, such as the blending of olive oil or wine while in storage to achieve certain finished product specifications.

Blending may also result from **idiosyncratic** design features of a particular processing line. Any time a process has, in-line, a tank (or any vessel with liquid capacity) into which a stream is fed continuously, and out of which a stream is taken continuously, blending of many product codes will usually occur at that point. One example is recycling product away from the filler to an earlier point (or tank) in the process and using that same tank to feed the filler continuously. Another example is the use of surge capacity (such as an aseptic surge tank) to keep a continuous process running continuously when it is fed with batch-wise inputs. Both examples are likely to cause mixing of many product codes.[9]

Idiosyncratic blending can be especially troubling for traceability systems since it might affect a large amount of production and many **lot** codes.

Idiosyncratic blending usually requires a fluid medium in a continuous process. Most batch and dry processes are generally free of **idiosyncratic** blending.

Any blending will necessarily reduce the **precision** of the traceability system.

---

[9] About the only time when these examples would not mix lot codes is if the lot codes represented very long times.

All blending should be considered as transformations that will require a **record keeping step**. Keeping records of **purposeful** blending is easy enough since it results from conscious management decisions. Record keeping for **continuous** blending is only somewhat more difficult in that the blending takes place with the knowledge of the operators even though no decision was required for it to happen. **Continuous** blending requires only that more than a single **lot** number of a raw material be linked to a single **production code** of finished product. In many cases, **idiosyncratic** blending will present a record keeping problem. It takes place both without any decision by management and frequently without the explicit knowledge of the operator. Also, **idiosyncratic** blending can cause the mixing of many **lot** numbers.

In spite of its challenges, **idiosyncratic** blending can be handled in a way consistent with traceability through record keeping. However, it is **idiosyncratic** blending that can reduce the **precision** of a traceability system more than any other type of blending.

Besides the standard record keeping approach, **idiosyncratic** blending requires a review of the process to determine where it might occur and what effect it might have on the **precision** of the traceability system. This review should be documented and made a part of the traceability system.

9.4.2 Transformations in Processed Foods

Consistent **internal traceability** requires records be kept wherever there is a transformation of product from one form to another. In fresh produce, a transformation could be cooling, sorting, grading or any process, even storage, which may change the form or condition of the product. For processed foods, the definition of a transformation has the potential to expand greatly since there are more operations involved. It could be any operation in common with fresh produce plus any processing operation such as the addition of one or more **ingredients** to another, a fractionation (such as grading, evaporation, filtration, crystallization or extraction), or any processing step (such as sorting, washing, blanching, freezing, heating, cooling, or filling) that may cause a transformation in the product.[10]

[10] It should be mentioned that certain transformations (such as finished product storage) that may be important for fresh produce are not as important to processed foods. The reason lies in the difference in shelf life between the two product categories. Storing fresh produce for a week will likely have a major effect on its shelf life in the trade as well as its quality on release. This effect could be described as a "transformation". If most processed foods were stored under the same conditions for a week, the change in the product would not be measurable. No "transformation" would take place.

9.4.3 Transformations That May Not Require a Record-Keeping Step

Keeping traceability records at each point of the process may or may not be useful to a particular **processor**, depending on the objectives of the traceability program. It is easy to argue that almost every step in a process causes a transformation in the product and would require a **record keeping step**. For example, a drying step for filled, pasteurized and water-cooled containers is a transformation, but a **record keeping step** for traceability at this point will not be useful.

A more ambiguous example is the placing of individual cartons of frozen vegetables into a shipping case. It is a transformation, but keeping a record of this operation for traceability may have different utility for different **processors**:

- **Processor** A uses the same **production code** for the shipping case that appears on the individual cartons. He believes a **record keeping step** at this point is not important.
- **Processor** B may consider a **record keeping step** at this point to be desirable since it could help trace and implement corrective actions should anything go wrong in this operation that was not picked up in the trade or during production by routine checking.
- **Processor** C may be concerned that different **production codes** of cartons would end up in the same case almost every time the **production code** changed. This is the example of **continuous** blending, given above. This **processor** may want a **record keeping step** at this point to assure that both **production codes** appear on that one case that contains cartons of two **production codes**. The same argument could be made for a palletizing operation.
- **Processor** D may recognize that its traceability system lacks the **precision** that would make the overlapping of two **production codes** in a single shipping case important.

Likewise, one **processor** may consider a **record keeping step** for traceability at a continuous freezing operation to be unnecessary since a record of what entered the freezer exists and nothing is added or taken away from the product except heat and has, therefore, no bearing on traceability as such. Yet another **processor** may believe that traceability can be used for diagnosing problems in the trade that can be linked to a production step. For this **processor**, a **record keeping step** at the freezer may be desirable.

One illustration of a transformation which is not amenable to a record-keeping step is in the manufacture of olive oil. Olives are washed, pressed, heated, and centrifuged in one continuous and close-coupled operation. While a **record keeping step** is certainly appropriate for this series of operations, any attempt to keep separate records of a single operation will not be meaningful. These types of close-coupled continuous operations are common in food processing. The best that can be done with such an operation is an over-all material balance on each **lot** number entering the operation (kg raw material in, kg pomace, water and olive oil out).

9.4.4 Transformations That Require a Record-Keeping Step

As illustrated above, few rigid rules apply to which transformations require a **record keeping step**. Generally, a **processor** should consider only the following transformations to be in need of a **record keeping step** for traceability:

- The addition of an **ingredient** (such as batching)
- Fractionation of a product stream which produces more than one useable output (such as size grading, but not sorting or dehydration)
- Any operation that results (or might result) in blending of two or more **lot** numbers of raw materials or **ingredients** into a single finished **production code**.

9.4.5 Some Common Problems in Traceability

Implementing a traceability system will always have challenges (or problems), many of which are common to many **processors**.

9.4.6 Dealers and Other Middle-Men

Many **processors** over the world purchase a significant fraction of their raw produce from dealers. Dealers tend not to reveal the sources of their produce, the produce at times comes with no **lot** code, and dealers will at times target produce they know to be high in pesticides or heavy metals to companies they know do not test for these attributes frequently. The solution to the dealer problem is threefold:

- Keep it simple
  Remember that the **processor's** only responsibility may be **one up/one down**. If this is the case, recording the name of the dealer, the date, product description, and quantity may be sufficient.
- Assign batch codes to inputs where none exist
  This batch code may be any system to identify the dealer, product description, quantity and the date received. It will be necessary to track his produce through the process and to link it with the finished product.
- When this is not enough
  Many dealers do not keep records for traceability, and it is clear that traceability is almost always broken at the dealer level. If this situation is not acceptable to the **processor's** customers, then that **processor** may have to guarantee the hygienic quality of the various **lots** provided by the dealers by appropriately testing them. In this way, that **processor** becomes the first link in the traceability chain. If the **lots** of produce are too small for cost-effective testing, then perhaps the **processor** should cultivate larger dealers or consider not using that commodity altogether.

Dealers in most places represent reality, at least over the short to medium term. However, many **processors** are acquiring and expanding their own farmland to avoid dealers. Other **processors** are pushing the dealers to hold up their end of the traceability contract. Presently, the dealers have little incentive to do this and the situation will continue either until they start losing business to other dealers with traceability systems in place, or until all begin to feel the pinch of vertical integration in the processing industry.

9.4.7 Ingredients and Packaging Materials with No Production Code

Frequently, some **ingredients** and packaging materials come without batch codes. Salt, sugar, and locally-manufactured packaging materials are examples. The approach to this problem is similar to the problem of dealers:

- Keep it simple
  Remember that the **processor's** responsibility may be **one up/one down**. If this is the case, recording the name of the dealer, the date, product description and quantity may be enough.
- Assign batch codes to inputs where none exist
  These batch codes may use any system to identify the product, the date received, and the manufacturer. It will be necessary to track these inputs through the process and to link it with the finished product. Encourage suppliers to provide batch codes and certificates of analyses guaranteeing the hygienic quality of the product.
- When this is not enough
  Although the **processor** is carrying out its **one up/one down** responsibility by recording the supplier of an **ingredient**, any **ingredient** that comes without a **lot** number carries with it the probable risk of breaking the chain of traceability. Some **processors** may not be comfortable with this risk. The risk can be managed if the **processor** guarantees the hygienic quality of the un-coded **lot** by performing appropriate tests on each **lot** (i.e. heavy metals for salt, etc.). This may not be feasible for small **lots,** and any **processor** receiving such small **lots** should consider the risk and, if possible, utilize for export only those **lots** for which the hygienic quality can be guaranteed.

9.4.8 Idiosyncratic Blending

In most instances, this type of blending of product codes can not be tracked through the process. However, the **processor** needs to know where in the process it may occur and how many code numbers could be affected. For example, if some aseptic juice coded 2:00 p.m. was known to be contaminated and it came from a processing line with a surge tank prior to the filler then a decision should be made to evaluate all product from 2:00 p.m. to the next **CIP**.[11]

9.4.9 Mixing of Filled Containers in Process Equipment

Frequently, filled containers will undergo processing in such equipment as an exhaust box, pasteurizer/cooler or a retort sterilizer. Sometimes the containers are coded at this point, sometimes not. These operations, at times, mix product codes. Mixing product codes can be handled in one of two ways:

- Record the mixing
  Keep track of the mixed codes and accept the resultant loss of **precision** in the traceability system.
- Prevent the mixing
  Mixing of product codes in these types of equipment can easily be prevented through simple modification of the systems and through monitoring. Mixing of product codes should be prevented where possible.

9.4.10 How to Proceed, Generally

Below is a generic (non-sector specific) guide to how to approach the establishment of a traceability system.

- Determine what products and **ingredients** are to be included in the traceability system and how it would fit into the existing quality management system.
- Determine the objectives, target regulatory requirements (or other standards) and company policies that are to be met by the proposed traceability system.
- Draw a flowchart of the process, identifying each step and the documents/records/sheets currently used in each step.
- Determine what data should be collected from the suppliers, kept by the **processor** and shared with customers. This decision will depend on the **depth** (one-up, two-up) desired for the traceability system.
- Assign (when necessary) and record **lot identifier**s for incoming raw materials, **ingredients,** and primary packaging materials.
- Identify each step in the process that would require the assignment of a new batch code. Generally, these are steps in the process where either:
  - An **ingredient** is added (such as batching)
  - A product stream is fractionated and produces more than one useable output (such as size grading, but not sorting or evaporation)

[11] Since most aseptic lines operate continuously, the remaining level of contaminated product in this example will never be zero until the line is shut down and cleaned. In the case of idiosyncratic blending, the level of the contaminant over time can not be calculated, only inferred qualitatively.

  - The operation results (or might result) in blending of two or more **lot** numbers of raw materials or **ingredients** into a single finished product code.

- Compose a data sheet for each step identified above which would link the new batch code to its inputs.
- For each finished product, determine a procedure to assign a **production code** (which may include information from the **label**) that can be linked to **lot identifiers** of inputs as well as pertinent **internal traceability** data. It is this **production code** that will appear on the outside of every package of finished product. The **production code** should be applied to each package as soon as possible after sealing.

  - **Production codes** are usually based on the time of sealing. Typically, the first bit of information is the date of production.
  - The second piece of information is the time of day. This time may be expressed in any way that suits the traceability objectives. It may present the exact time of sealing (i.e. 0946) or a code (i.e. B) signifying an hour, a span of hours, or a shift.
  - The third piece of information might be a line or factory designator in the event that the same company makes (or contracts to be made) the same product on more than a single production line or at more than one location.
  - Be sure that the finished product code is linked to all inputs and any changes that they may have undergone in the factory. This linkage may be limited only to direct inputs and indirect inputs caused either by continuous or purposeful blending. Potential for idiosyncratic blending should be noted but not necessarily tracked through the system.

- Take care that all batching operations, primary and secondary, have batching sheets and that these sheets are linked to the finished product code.
- Link any relevant food safety data to incoming materials and finished products.
- Determine the protocol for data retrieval (where data are stored, how long they are retained, who has access to the information, etc.).
- Assign responsibilities within the organization for the various parts of the traceability system.
- Develop and implement a training program for these individuals.
- Develop a system of auditing the traceability system, recording the results, and assessing corrective actions.

9.5 Conclusions

Many traceability models applied to processed foods are based on fresh produce. However, processed foods are fundamentally different from fresh produce because of their potential to blend input **lots** within the process. Ignoring this blending will likely break the chain of traceability within a factory that otherwise keeps good records of inputs and outputs.

Glossary of Terms

Breadth The amount of information that a traceability system demands,

CIP Cleaning in-place

Common carrier The entity, usually a third party, responsible for delivery of finished product

Depth How far back or forward a traceability system traces or tracks (such as one up/one down)

HACCP Hazard analysis and critical control point

Ingredient Any substance, including water, intentionally incorporated into the food during its manufacture, preparation, or treatment

Internal traceability The ability to track what happens to raw materials, ingredients, primary packages, and finished products inside the processor's operation

ISO International Organization for Standardization

Label A written statement appearing on each container of processed food. Typically, a label contains, at a minimum, the product name, form & style, its net contents and the identity of the manufacturer. The information contained on a label does not change on a regular basis.

Lot A collection of materials (ingredients, raw materials, or primary packaging materials) produced in the same time period under conditions as nearly uniform as possible, designated by a common code for identification

Lot identifier A code which uniquely identifies a lot. Terms such as batch number, batch code, and lot number may all be synonymous with lot identifier. A production code and information from the label may also be considered to be a lot identifier.

One up/one down A description of one level of depth of a traceability system. In a one up/one down system, each participant in the food supply chain is responsible for maintaining records about the products they receive, their use (i.e. the link between inputs and outputs) and to whom they were shipped, or sold.

Precision The degree of assurance that a traceability system can pinpoint the movements of a particular ingredient to a single lot of finished product, or conversely, can pinpoint all the lot(s) of raw materials, ingredients, or primary packaging materials that make up a finished product

Primary producer A farmer or grower

Processor A member of the supply chain that typically receives inputs from primary producers, suppliers of ingredients and packaging materials and/or common carriers and transforms these inputs into some other form. This other form is typically packaged in a way that would preclude the addition of more inputs or processing without opening it. A supply chain may have more than one processor.

Production code The identification printed on each container of processed food at the time of its manufacture. The production code uniquely identifies when the product was packaged (the year, day, and time period). It also uniquely identifies, by examination of the factory's records, which factory manufactured the product[12] and what production codes of raw agricultural commodities, ingredients, and packaging materials went into the product. The production code should be changed with sufficient frequency to enable ready identification of lots during their sale and distribution. Production codes should not extend over a period of more than one personnel shift.

Record keeping step Also known as a critical tracking event,[13] a point in a process where records are taken and new batch codes assigned for the purpose of traceability. Data taken could include input lot numbers, times, and other data to link the inputs to the outputs.

Ultimate processor The last processor in a value chain to add ingredients or otherwise process a product

Suggestions for Further Reading

Annotated EC Guidance on the Implementation of articles 11, 12, 16, 17, 18,19 and 20 of Regulation (EC) N° 178/2002 on General Food Law Conclusions of the Standing Committee on the Food Chain and Animal Health. http://www.foodlaw.rdg.ac.uk/pdf/eu-05007-food-law-guidance.pdf

Can-Trace Decision Support System for Food Traceability Agriculture and Agri-Food Canada, 2004. http://www.can-trace.org/portals/0/docs/Can-TraceDecisionSupportSystemforFoodTraceability.pdf

Can-Trace Integration Guidelines Final Report April 2006 Agriculture and Agri-Food Canada. http://www.can-trace.org/portals/0/docs/Can-Trace%20Integration%20Final%20Report%20April%202006%20-%20mjf.pdf

Can-Trace Multi-Ingredient Working Group Multi-Ingredient White Paper, March 2006 Agriculture and Agri-Food Canada. http://www.can-trace.org/portals/0/docs/Multi%20Ingredient%20Final%20Report%20March%202006%20-%20mjf.pdf

Can-Trace Produce Pilot Project Report Agriculture and Agri-Food Canada, 2004. http://www.can-trace.org/portals/0/docs/Can-TraceProducePilotProjectReport.pdf

Can-Trace Technology Guidelines Agriculture and Agri-Food Canada, March 2006. http://www.can-trace.org/portals/0/docs/Can-Trace%20Technology%20Guidelines%20Mar%202006%20-%20mjf.pdf

CODEX: principles for traceability/product tracing as a tool within a food inspection and certification system CAC/GL 60-2006. http://www.codexalimentarius.net/download/standards/10603/CXG_060e.pdf

Food Traceability Report (weekly e-newsletter). http://www.foodtraceabilityreport.com/home.asp

Ian Smith, Anthony Furness (eds) Improving traceability in food processing and distribution. (CRC – 31 Mar, 2006)

---

[12] This information is needed in the production code only in the event that multiple production sites exist for the same product for the same company. Otherwise the product's label, with the identity and address of the company, will take care of this item. It follows then that the production code can include information that is presented on the label.

[13] http://www.ift.org/cms/?pid=1002160

International Standard ISO 22005, First Edition, 2007-05/15: traceability in the feed and food chain-General principles and basic requirements for system design and implementation
Report on Can-Trace National Food Traceability Consultation Sessions Agriculture and Agri-Food Canada, June 2005. http://www.can-trace.org/portals/0/docs/ConsultationSessionReport.pdf
Report of the Can-Trace Small and Medium Enterprises (SME) Working Group Agriculture and Agri-Food Canada, 19 May 2004. http://www.can-trace.org/portals/0/docs/ReportoftheCan-TraceSMEWorkingGroup.pdf
Traceability Decision Support Tool Agriculture and Agri-Food Canada. http://www.can-trace.org/portals/0/docs/Can-Trace%20Decision%20Support%20Template%20v1.03.xls

Chapter 10
Availability of Shelf-Stable Foods: Advances in Production of Meal Ready to Eat (MRE)

Magdy Hefnawy

Abstract The global food supply chain continues to be vulnerable to threats from a variety of directions. Food security and safety became a hot topic worldwide in the academic research arena and food industry practices. Recognizing that food could be a cause of significant disaster in communities, the need to secure safe and wholesome food is imperative during and after terrorist acts in addition to natural and/or man-made disasters.

The availability of foods ready to eat which victims can consume immediately during and after catastrophic events is considered necessary. Foods in retort pouch in form of Meal Ready to Eat (MRE) could serve the ideal food during crises. The history of the development of retort pouch goes back to Second World War and has not become a reality until the 1980s. Food in retort pouch has proven to be a successful combat ration and the food of choice for military during conflicts and wars.

Production capabilities and availability of foods in retort pouch are evolving. The ability of existing food processing methods and advanced/emerging technologies to produce safe food in retort pouch has proven its reliability and conformance to food regulations. Currently, foods in retort pouch are existed in every supermarket and in sufficient supply of nutritionally adequate and safe for humanitarian purposes in peacetime or during crises.

Keywords Meal ready to eat • MRE • Production • Retortable foods • Shelf-stable foods

10.1 Introduction

Disasters whatever the cause natural or man made can strike suddenly without warning. In recent years, several natural disasters as well as man made threats have stricken the world. When natural dissipaters happened, such as tsunami in Asia,

---

M. Hefnawy (✉)
Ag-Tech International, Inc., 237 Cal Dobson Trail, Greeneville, TN 37743, USA
e-mail: magdyh@centurylink.net

M. Hefnawy (ed.), *Advances in Food Protection*, NATO Science for Peace and Security Series A: Chemistry and Biology 1, DOI 10.1007/978-94-007-1100-6_10,

Katrina in the US, or earthquake in Haiti required fast response to provide the victims with their basic life need: FOODS. The type of foods, how fast, and in what shape should be delivered to the victims make a great difference that should influence the industry and governments preparedness. Shelf-stable foods that are ready to consume and provide the nutritional requirements for different ages, cultures, and religions need to be available and ready to be delivered to the victims in timely mannered [1]. Aspects in forms and formulations of shelf stable foods to meet the need during or after disaster in different regions should be evaluated. Industrial and government planning and preparedness are the key in swift and effective response to any disaster and definitely avoiding food crises.

Shelf-stable foods lend itself for the purpose of avoiding food crises due to natural disasters or terrorist attacks. Such food crises that could jeopardize food safety and/or food supply may be minimized with the availability of shelf-stable foods. The technological advances in producing shelf-stable foods in the form of Meal Ready to Eat "MRE" are reviewed in this chapter. The progress in the retort pouch technology and the preservation of foods in retort pouch related to pouch materials, filling, sealing, and commercial sterilization process helped the availability of almost any food in good quality [2].

10.2 Historical Development of Foods in Retort Pouch

In 1940 Wilber Gould and his research associates, at Ohio State University reported the first recorded idea of proposing a flexible packaging material for thermal processed foods.

The next recorded study was by Hu et al. [3] and Nelson et al. [4] at the University of Illinois. These researchers did the first systematic screening of all available films for permeability to water vapor and oxygen, resistance to boiling water and steam at 250°F. They found two acceptable materials (polyester or Mylar, and trithene) and then prepared low-acid test packs. They also applied and refined the commonly used water process technique with super-imposed overriding air pressure to prevent pouch bursting during cook.

Frank Rubinate 1957, military researcher foresaw the functional advantages of the retort pouch for improving combat rations.

Keller 1959 [5], conducted an extremely thorough investigation of films, applications, retort processes, and the use of paperboard folders as secondary packaging.

Beadle, 1959 established the reliability of heat seals. He, also, established the seal strength advantages of a round jaw heat sealer over a flat bar.

In the year 1960, Star Di Agrate Brianza (Star Foods) of Italy initiated limited production of retort-pouched foods commercially.

In 1961, the military researchers made commercial test packs of at least seven food items and ran field abuse tests to prove durability of the pouch package concept.

Floren Long 1961, of Continental Can reported on many varied laminations being tested and finally offered a polyester/aluminum foil/ polyolefin laminate as the most successful combination of materials.

The first commercially recognized market introduction of the retort pouch occurred in 1967, when Otto Nielsen provided a variety of sausages and sliced meats sold in England under several brand names, including Swift.

In 1967, Specialty Sea Foods of Washington was the first company in the U.S. to commercially use flexible packaging for retorted foods.

In 1968, Toyo Seikan of Japan started commercial introduction of retort pouched foods in Japan. The lack of a competitive line of frozen foods, a desire for high quality foods, and an acceptance of the labor-oriented, slow-speed production lines as a challenge – not a roadblock – turned the retort pouch into a more than 500 million packages a year consumer market in Japan [6].

1969 was a year of historical significance for the retort pouch when N.A.S.A. selected a number of foods sterilized in retort pouches for the Apollo 11 moon orbit mission. Up until this point, all foods had been dehydrated and freeze-dried.

In 1972, the U.S. Army Natick Laboratories completed a study that involved packaging specialists throughout the food industry to determine the reliability of a production line to manufacture a variety of food product types with minimum acceptable reject rates [7].

In 1973, Swan Valley Foods of Canada marketed the first commercial retort products in North America. Other companies experimenting and seriously evaluating the retort pouch in the early seventies included Kraft, Hormel, Green Giant, Hunt-Wesson, and Libby's among others.

"After a brief span of 52 years, retort pouch development came to a complete stop. Based on some research studies indicating a possible migration of carcinogenic compounds from the pouch laminate adhesives to the food during the heat sterilization process at elevated temperatures of 240–250°F, prior approvals by the FDA and USDA were withdrawn on all retort pouch materials," Omega Research Associates 1992 [8].

In May 1977, the FDA gave full approval of retort pouch and a month later the USDA dropped its barriers.

Immediately after the FDA approval in 1977, three major suppliers – Reynolds Metals, Continental Can, and American Can developed alternate heat-bonded laminates acceptable for foods and retort heat process (Continental's group has since become Ludlow Flexible).

In October 1977, the ITT- Continental Baking Co. introduced in several test cities the first U.S. commercial retort pouch. This test was very successful, but ended in 1978 after 6 months due to inadequate production lines.

In 1978, the U.S. government opened up with solicitation for 24 million called "MRE" or Meal, Ready to Eat, as opposed to its predecessor, the "MCI" or (Meal, Combat, Individual). Contracts were awarded in 1979 and continue to be manufactured today. The military market indicated a commitment to the retort pouch concept, and also offered potential new processors a learning experience opportunity.

The progress in the retort pouch technology allowed the US Military to replaced Meal Combat Individual(MCI) with Meal Ready To Eat (MRE).The use of foods in retort pouch instead of cans offered greater variety items that received higher acceptability ratings, and lowered cost of meals [9].

10.2.1 Benefits of Foods in Retort Pouch

The pouch combines the desired properties of most current commercial food systems. The retort pouch achieves the same shelf stability of metal and glass containers, less brine, syrup, or gravy has to be added, and the sensory quality is closer to that of frozen foods rather than canned foods because of the shorter thermal process time required to achieve commercial sterility. The pouch, also, has the convenience of a cook-in pouch since the product can be heated to serving temperature in 3–5 min as compared to the 10–20 min needed for frozen boil-in-the-bag pouches [10].

Shelf-life. Shelf life of food in retort pouch is at least equal to that of foods in metal cans. It does not require refrigerating or freezing. Food stored in a retort pouch can be stored in climate controlled warehouses until needed. This way, the MREs can be kept for about 10 years, before being used [11].

Convenience. Opening the pouch requires only tearing the package across the top at the notch in the side seam. Food doesn't require additional cooking or preparation, however, if need to be heated to serving temperatures, can be accomplished in few minutes by immersing the pouch in boiling water or heated directly on top of hot surface. Features for consumer convenience can be added to the pouch such as zip lock and spouts.

They are safer than cans or glass jars which could cause injury from cut on the metal can or broken glass.

Storage and disposal space. Storage space of the retort pouch is much less than that for cans. Disposal space is also less.

Taste/nutritional value. Reduced heat exposure results in improvements in taste, color, and flavor. There are also less nutrient losses, and with controlled fluids, nutrient fortification is possible. The product around the periphery is not overcooked, as it may be with cans, and the quality should be better as the product is truer in color, fresher in favour, and firmer in texture,

Other benefits. The retort pouches lend themselves to portion control and thus have a consumer advantage for single people and the elderly.

Pouch contents can be completely squeezed out, whereas as much as 5% of the contents typically remain in the can.

There are also benefits for the food manufactures that producing food in retort pouch should provide:

Empty pouch storage space and weight. Empty retort pouches offer processors a reduction in storage space and lighter weight. Compared with empty cans, an equal number of retort pouches use 85% less space. One thousand 4.5×7 in. pouches weigh 9.4 lb compared with one thousand 211×304 metal cans weighing 112.5 lb. A 45-ft trailer holds less than 200,000 eight-ounce empty metal cans, whereas on the same trailer over 2.3 million empty preformed pouches can be shipped.

Label display space. In comparing retort pouches and metal cans of the same fill size, the display area advantage is greater with retort pouches.

Package energy intensities. The pouch is the least energy intensive of the packaging systems.

The amount of the heat energy required for thermal processing is less since the container thickness and the amount of packing medium will be reduced, resulting in a more rapid rate of heat penetration.

Energy required to produce four types of 8 oz. food containers

Container	BTU's/container
1. Retort pouch	
Polyester, aluminum foil, adhesive, polypropylene, inks, carton	1,934
2. Frozen food	
Aluminum container, coatings, plug lids, carton	2,819
3. Glass jar	
Lid (steel), seal compound, label, glue	3,174
4. Can (211 × 300)	
Steel, tin, coatings, label, glue	3,560

Source: Omega Research Associates.

External corrosion. Pouches do not corrode externally, and there is a minimum of product-container interaction.

Marketing opportunities. The retort pouch can be merchandised anywhere in the store while frozen are limited to the freezer and cans are not as versatile [12].

10.3 Production of Foods in Retort Pouches

The production of thermo-processed foods in retort pouch is similar to the production of foods in cans or glass jars. The fundamentals of producing commercially sterilized foods are applied and are discussed as follows:

1. Food Products
2. Product Preparation
3. Pouch Materials
4. Pouch Filling
5. Air Removal
6. Pouch Sealing
7. Packaging Machines
8. Retort Racks
9. Retorts

10.3.1 Food Products

Retort pouches are applicable and advantageous to a wide variety of foods. Retort pouches can replace metal cans or glass/plastic jars. Retort pouch is feasible for almost any food product with the exception of food with sharp bones, shells, or other sharp edge components. However, it has been reported that Coq au Vin with complete half chicken has been successfully produced in retort pouch [13].

Retort pouch lend itself to improving quality for shelf stable products such as [14]:

- Soups
- Sauces
- Gravies/sauces containing particulates as Meats, Poultry, Seafood, Vegetables and/or Pasta
- Vegetables
- Fruits
- Bakery
- Meats/Poultry/Seafood with or without fluids

10.3.2 Product Preparation

Product preparation prior to filling into pouches is similar to metal / glass canning. Batch size is determined with retort capacity. Filled pouches should be retorted within 2 h from filling. Weighing ingredients, blending and cooking should be carried out in compliance with local and government regulations [15].

10.3.3 Retort Pouch Requirements

10.3.3.1 Retort Pouch Materials

The basic requirements for retort pouch film related to its functionality as well as resistance to handling abuse, and most important is resistance to the thermal processing (248–270°F): The interaction between the food product and the film during the sterilization process is important in maintaining the quality of the food and the integrity of the pouch. The following are the basic requirements of the pouch material:

- Permeability
  - Low gas (O_2) permeability less than 1 cm^3/100 in.^{2}/24 h at atmosphere
  - Low water vapours transmission rate less than 0.05 g/100 in.^{2}/24 h at atmosphere

- Resistance to temperatures
  - From 0°F to 260°F to withstand storage condition and sterilization process
- Comply with FDA regulations
  - Resistance to penetration by food components, and migration of film components to foods.
- Physical strength
  - Resistance to any handling abuse during manufacturing and during distribution channels

10.3.3.2 Retort Pouch Structures

There are many films have developed by several suppliers that meet the above requirements. Material developments in retortable films are improving the bond strength using gamma radiation at six megarads without adverse effects on the functionality of the pouch [16]. Currently, the following material structures are being used in the manufacturing of retort pouches:

2 - ply (polyester or nylon/polypropylene)
3 - ply (polyester/foil/polypropylene)
4 - ply (polyester/foil/polyester/polypropylene)

The structure of the film determines the shelf life of the product, Shelf-life minimum of 12 months (low barrier) to 3 years (foil layer).

The US military adopted the 4-Ply film (Quad laminate film) for the use for MRE combat ration program (Fig. 10.1):

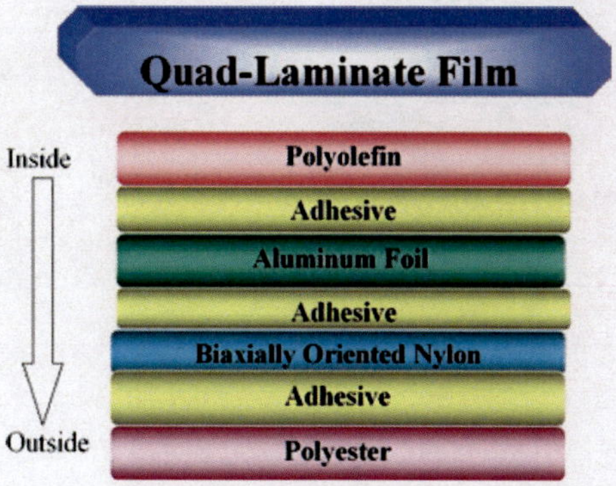

Fig. 10.1 Military pouch, film structure

10.3.3.3 Retort Pouch Types and Designs

The design, shape, and type of the retort pouch is manufactured to accommodate the food type, the required serving size, and processing requirements of the manufacturer. The pouch could be made of clear or opaque film, and foil laminate (higher barrier) for long shelf life. It can be deigned for stand up (gusseted) or flat (pillow) pouch (Fig. 10.2).

10.3.3.4 Retort Pouch Features

Due to the flexibility nature of the film structure and shapes of the pouch, several features can be introduced to the pouch [17]. The type of the food product and end-user preference dictate the following features to be used in the pouch:

⇒ Spout	⇒ Zippers
⇒ Notches	⇒ Peg hole

Fig. 10.2 Different shapes and sizes of retort pouches

10.3.4 Pouch Filling

The filling system is selected based on the type of food. Foods may be divided to two categories based on the way product moves to the filling equipments:

1. Pump-able products	2. Place-able products
Product moves with pump:	Product contains big pieces:
Sauces	Steaks
Small particulates /gravy	Patties
Liquids (thick or thin)	Chicken/fish fillet
Dry products	

Deferent types of pumps are available and used successfully with filling retort pouches including positive displacement pumps, gear pumps, and piston type pumps. The dosage is controlled either volumetric or by gravimetric (weight).

10.3.5 Air Removal

The removal of air in pouches before sealing will prevent pouch distortion or breakage during the sterilization process. The heat transfer in the pouch is more uniform and predictable when there is no air in the pouch [18]. The product quality is affected with the presence of oxygen and product stability is better with the absence of air in the pouch. Several methods to remove air from the pouch before sealing are used:

- Steam Flush
  After pouch is filled with product, steam is injected into the open pouch immediately before sealing.
- Vacuum
  - Snorkel
    A suction tube is inserted in the filled pouch to remove the air and pulled out immediately before sealing.
  - Chamber
    The filled pouch is positioned in a vacuum chamber, controlled the air withdrawal, and sealed under vacuum.
- External Pressure
  This method is used for fluid products where the air is removed by the physical displacement caused by squeezing action of both sides of the pouch to rise the content of the pouch just up to the seal area.

10.3.6 Pouch Sealing

The most important factors that determine the integrity and functionality of a pouch are film structure and pouch seals. A pouch seal has been a persistent concern in the quality of the retort pouch. Two methods are being commonly used:

- Heating bars
  Filled pouches are sealed with heated bars, usually covered with Teflon to prevent plastic contamination of the bars. Often, three sets of bars are used, first set to preheat the sealing area, second set is heated to bond the inner layers of the pouch, the third set is a cold bar station where the sealed pouch is pressurized by a set of cold bars to set the seal [19]. Heating and cooling may be achieved with one set of sealing bars at one station (Fig. 10.3). (Impulse Bar).
  Critical factors in heat-sealing the retort pouch include:

  – Seal bar temperature
  – Pressure exerted on the seal by the sealing bars
  – Dwell time (time seal bar pressure is exerted on seal)

- Ultrasonic seal
  The ultrasonic sealing of the retort pouch has proven to be stronger, reliable, and more economical than heating bars [20]. Sealing is based on the conversion of mechanical sound waves into frictional heat resulting in melting the inner surface of the pouch. The applicable Ultrasonic frequency for sealing the retort pouch materials rang between 20 KHz. and 70 KHz. The ultrasonic introduces the vibration energy directly into the package materials and the resulted friction

Courtesy of Toyo-Jedoki Co.

Fig. 10.3 Three stations, heating bars: preheat–heat–cool (Courtesy of Toyo-Jedoki Co.)

Courtesy of Hermann Ultrasonics

Fig. 10.4 Ultrasonic sealer (Courtesy of Hermann Ultrasonics)

heats the inner layers of the package that bond with applied pressure. The advantages of the ultrasonic seal may be summarized as follows (Fig. 10.4):

- Heat generated at interface by vibration
- Inside temperature is higher than outside
- Mechanical hammering displace contaminant

10.3.7 Packaging Machines

Packaging machine selection is based on the operation capabilities i.e. Type of product, production capacity, and produced pouch sizes

- Form fill seal packaging machines:
  - Horizontal form fill:
  - In the "form/fill/seal" operation [21], where a multi-layered laminated web of polyester/polypropylene/aluminium foil is run along a horizontal plane and moulded into concave (bowl) shapes. The pouches are then filled and a continuous web of multi-layer plastic is fed from an overhead roller on top of the filled pouches. The top web is then heat sealed onto the pouches by heat sealing bars that descend from above. A vacuum is pulled on each pouch just prior to sealing the two material webs. After sealing, the individual pouches are cut from the web by a cutter wheel as the web exits the vacuum heat sealer (Fig. 10.5).

Fig. 10.5 Horizontal form-fill seal packaging (Courtesy of Multivac, Inc.)

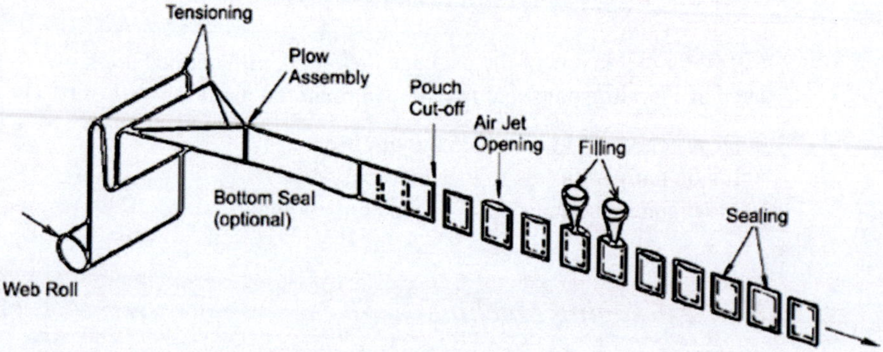

Fig. 10.6 Schematic diagram of vertical form fill seal

– Vertical Form-Fill-Seal machines also are available [22] (Figs. 10.6 and 10.7):

- Fill seal packaging machines
  Several types of fill seal machines are available in the market to produce food in retort pouch. Preformed pouch in different sizes, shapes, flat, gusseted, and features are also available. Producer may choose to start with roll-stock film, forming pouch machine can be attached to the fill seal packaging. The following are some examples of the existing types of fill seal packaging machines:

  (a) Rotary pouch filling machines (Figs. 10.8 and 10.9):
  (b) Linear filling sealing packaging machines (Fig. 10.10):

Courtesy of Cryovac Co.

Fig. 10.7 Vertical form-fill seal packaging (Institutional) (Courtesy of Cryovac Co.)

Fig. 10.8 Toyo Jedoki Model TL- AX1 (Courtesy of Toyo Jedoki Co.)

10.3.8 Retort Racks

Retort racks are available in stainless steel or plastic structures and vertical-slotted or horizontal racks. All designs feature perforations to allow efficient heat media exposure and circulation. The selected rack should ensure the pouch orientation in

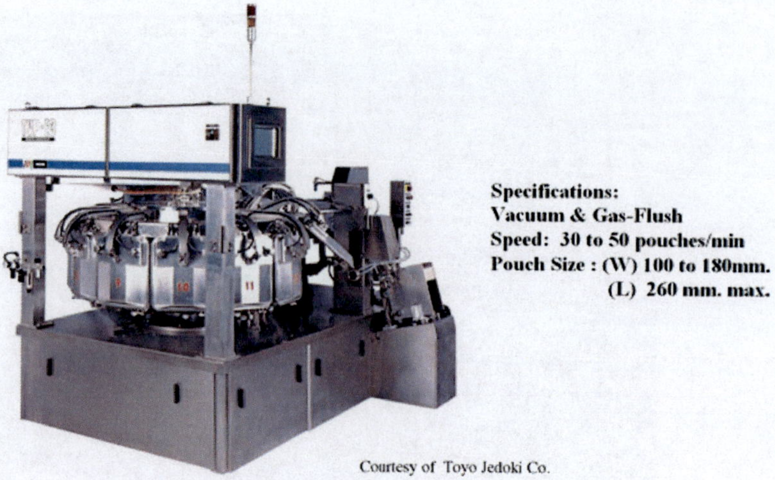

Fig. 10.9 Toyo Jedoki Model TVP E-3 (Courtesy of Toyo Jedoki Co.)

Fig. 10.10 Bartelt filling sealing machine (Courtesy of SOPAKCO Packaging, SC)

the retort should be confined to same maximum pouch thickness [23, 24] (Figs. 10.11 and 10.12).

10.3.9 Retorts

Retort or autoclave is the vessel where food is sterilized by heat in a pouch. Retorts are agitating/ rotating or non-agitating./still [25]. The food in retort pouch is sterilized in the retort under "full water immersion" or using steam-water spray (Figs. 10.13 and 10.14).

Fig. 10.11 Retort racks

Courtesy of STOCK America

Fig. 10.12 Multi purpose retort racks Model 1400-MP (Four permanent lanes can be divided into 7, 6, 3, or 2 compartments each) (Courtesy of STOCK America)

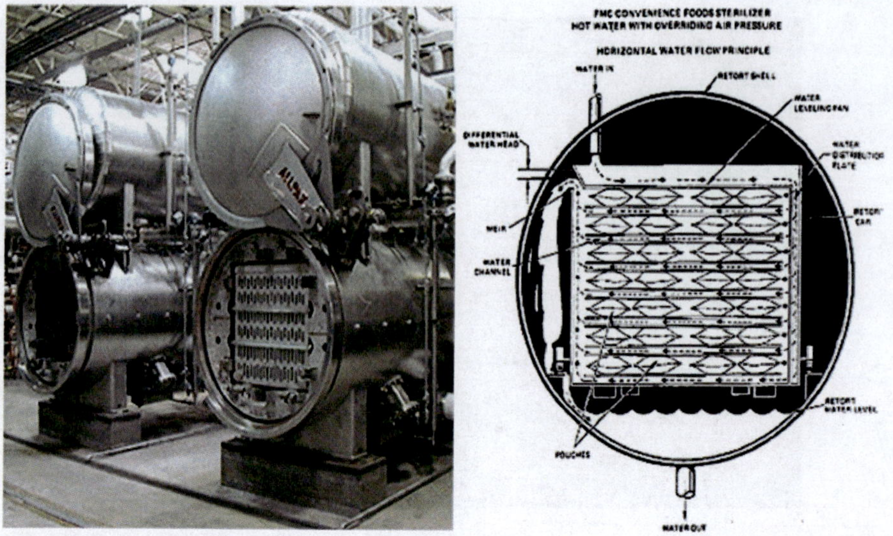

Fig. 10.13 Water immerse retorts (Courtesy of Allpax Co.)

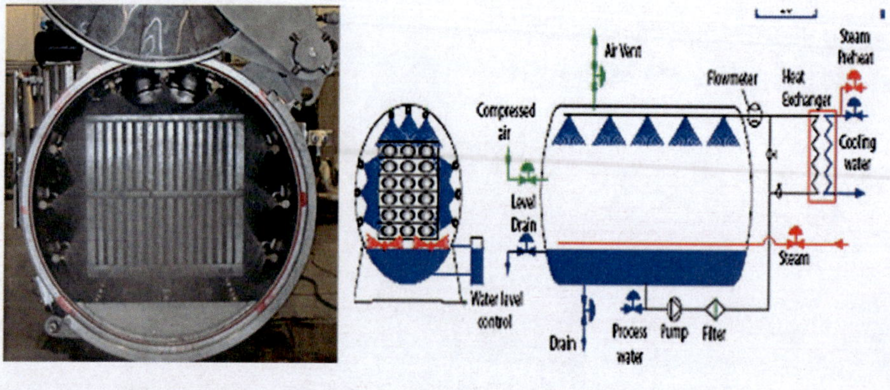

Fig. 10.14 Spray retort (Courtesy of Allpax Co.)

10.4 Principles of Thermal Processing

10.4.1 Basic of Thermal Processes

The knowledge of food preservation using heat started in late 1700s when Nicolas Appert solved the difficulties of feeding Napoleon's troops [26].

The thermal processing conditions, temperature and time, for pouches are determined by the same methods proven to be acceptable for metal cans and glass jars. The principle for establishment of thermal processes for canned foods is based on knowledge of food microbiology and of processing methods. The most easily regulated feature of the microorganisms' growth is temperature, although other sterilization agents may be used such as chemicals or high-energy radiation. Food contaminated with microorganisms, when placed in a sealed container, can be prevented from spoiling when subjected to high temperature for a sufficient time. The high temperature kills the microorganisms already present and the sealed container prevents recontamination of the food.

Determining the proper temperature and length of time needed to sterilize food in pouch utilize the standard concepts of F_o, D, and z values related to the microbial survival aspects.

F / F_o = "**F**" The time in minutes required to destroy a number of organisms with a known z value at temperature T. "**F_o**" = indicate the z = 18°F and the temperature is 250°F.

D = "**D_T**" The decimal reduction time. The time required to reduce bacterial population by 90% at T temperature

Z = "**z**" The number of °F required for the thermal resistance curve to traverse one log cycle.

"Generally F_o value suitable for commercially canned products are adequate for retort pouches, Lampi, 1979" [27]. The procedure for process determination is not simple. It depends upon the knowledge of a number of factors, including:

(a) the nature of the food product,
(b) the dimensions of the pouch in which it is to be packed, and
(c) the details of the thermal processing procedures used. It is imperative to gain information of the growth and survival characteristics and thermal resistance of the contaminating microorganisms. Once this information has been obtained, the process schedule is designed to achieve commercial sterility. The process parameters of temperature and time should be specified for a particular product in a particular container in order to destroy the spoilage organisms.

The regulatory agencies require that a "processing authority" having expert knowledge of thermal processing requirements establish scheduled processes. A processing authority is a person or organization having expert knowledge of thermal processing requirements for foods packed in hermetically sealed containers and having adequate facilities to make process determinations.

10.4.2 Process Schedule

The scheduled process or process schedule is designed by a processing authority to provide a commercially sterile or shelf-stable product. Commercial sterility (FDA) or shelf stability (USDA) means the condition achieved in a product by the applications of heat to render the product free of microorganisms capable of reproducing in the food at normal non–refrigerated conditions of storage and distribution.

The scheduled process includes thermal process parameters such as product initial temperature, process temperature and process time, plus critical factors may include any characteristic, condition or aspect of a product, container, preparation procedure or processing system that affect the scheduled process.

10.4.3 Establishment of the Thermal Process

Determining the proper temperature and length of time needed to commercially sterilize a hermitically sealed pouch depends upon the knowledge of:

1. The nature of the food product
2. The dimensions of the retort pouch
3. The thermal processing procedures used.
4. Knowledge of the growth and survival characteristics and thermal resistance of contaminating microorganisms

Based on the above information, the processes to sterilize a product can be determined. [Time and temperature required for destroying the spoilage organisms.] The establishment of a thermal process is based on two factors:

- First, the heat resistance of microorganisms – the amount of heat required for their destruction – must be known in each specific product.
- Second, the heating rate of a specific product must be determined. These two factors are used to calculate the scheduled process.

Once a process has been established for a particular food product, it is specific for that one product i.e. its formulation, its method of preparation, the container size in which it is processed and the kind of thermal processing system used. A process must not be altered unless specific instructions for change are obtained from a processing authority.

10.4.4 Heat Resistance of Microorganisms

The thermal resistance of microorganisms is dependent upon a number of factors that must be considered:

1. The growth characteristics of the microorganisms
2. The nature of the food

10.4.5 Factors Affecting Growth of Microorganisms

- pH
- Moisture Content
- Nutrient Content

- Oxygen Supply [Aerobes (oxygen); Anaerobes (no oxygen)]
- Temperature [psychrophiles 58–68; mesophiles 86–98; thermophiles 122–140 F]
- Water Activity A_w [<0.91 most Bacteria; <0.88 Yeasts; <0.80 Molds]
- Inhibitors [Chemical (Nitrite/Nitrate, sulphur dioxide); Natural (sugar, salt, organic acids)]

10.4.6 Product Heating Data Determinations

The rate at which a product heats can be measured using devices that monitor the change of temperature as the food is being heated. The determinations are accomplished with a temperature sensor or thermocouple located in the product at the slowest hating region of the container. The slowest heating region of the container will depend on the type of product, container size and processing method. The time/temperature data are monitored on a temperature recording device. These data are necessary to determine the heating rate of the product and to calculate the scheduled process. This procedure may be used for both low-acid and acidified products [28].

10.4.7 Heat Penetration Protocol

- Proper Identification of all parameters
  - Date of test
  - Pouch Set-up
    Punch hole in pouch
    Insert Packing Gland with Gasket
    Thermocouple (TC) through Packing Gland
    Insert TC tip into particulate (meat) or packed material (plastic tube or cheesecloth).
    Fill pouch to desired weight. Keep seals as clean as possible.
      1 oz. above Net Weight (within filler capabilities) or
      10% above Net Weight (within filler capabilities)
      Extra particulate ratio (determine worst case)
    Seal pouch
  - Product
    Name
    Formulation I.D.
    Conduction (slice meat)
    Viscosity
    Convection (gravies)
    pH
    Initial Temperature (IT)

Frozen (drill, or insert TC tip, seal, and re-freeze)
Largest particulate size (weight and/or dimension)
- Retort
  Identification #
  Style
  Temperature
  Pressure
  Rotation (RPM when applicable)
- Container information
  Construction
  Size [L x W x H (gusset)]
- Protocol
  Thermocouple (TC) Identification #, by container
  TC position (pre-retort)
  TC position (post-retort)
  Container position in retort
  Fill Weight (pre-retort)
  Drained Weight (post retort)
  Drained Weight : Net Weight Ratio
  Residual air (cc)
  Particle Size (L x W x H)
  Particle Weight (grams)
  PH
  Viscosity

10.4.8 Comments

- Basic principles that apply to heat processing of product in flexible containers are the same as the ones that apply to products in cans.
- Flexible containers make it difficult to keep TCs motionless and in consistent location. Must do multiple tests and use heating factors from slowest heating pouch scenario.
- Rotation and/or agitation of pouches may be undesirable due to delamination or abrasions.

10.4.9 Data Collection

- CaLSoft/CalPlex
  - "TechniCAL's CALPlex™ datalogger instrument is designed specifically for conducting heat penetration and temperature distribution tests [29]. It provides inputs for recording up to 32 temperature channels directly onto a

computer storage disk with the use of the CALSoft software and a laptop computer." *TechniCal Company.*

- DataTrace:
  - "The DataTrace systems are self-contained, wireless, high precision, data loggers that are used in critical manufacturing, quality control and transportation applications. They are used to measure temperature, humidity, and pressure inside a process or inside a product during manufacturing. In addition, the DataTrace data logging systems can be used to validate the proper operation of laboratory or manufacturing equipment either during its installation or for annual re-certifications. The product line consists of PC interface software, various accessories and individual data loggers such as temperature data loggers, humidity data loggers and pressure data loggers." *Mesa Laboratories, Inc* [30].

10.4.10 Heat Distribution in Retort

10.4.10.1 Temperature Distribution Protocol

Wiring Harness
Check Condition
Calibration next to Mercury-in-Glass (MIG) thermometer
Determine desired location of Thermocouple (TC) tips
Load Retort with Ballast and TCs in desired location
Draw an accurate "map" of TC locations

- Proper Identification of all parameters

Date of Test
Location of Test
Retort Identification
Person Conducting Test
Container Information
  Size
  Physical Arrangement
    Number of containers per layer
    Number of layers per cart
    Number of carts per retort
    Total number of containers

  - Test conditions

  Ballast (product)
  Initial Temperature (IT)
  Steam Header Pressure at beginning of test
  Steam Header Pressure during come-up
  RPM, when applicable

- Record Data
  MIG
  Chart
  PSIG at Controller (during CUT)
  PSIG at Header (during CUT)
    At pre-determined intervals (usually 1 min)
    Time Mercury-in-Glass (MIG) thermometer is up to set point
    Time all TC leads are up to set point
    Retort Temperature

10.4.11 Process Calculation

The processing authority will calculate a process using the product heating data and the thermal resistance data for the significant spoilage organisms or organisms of public health consequence expected to be present in the product. A number of ways exist to calculate the process.

10.4.12 Thermal Processing Considerations

Process design: Principles used in the development and application of thermal processes for pouched foods are the same as for conventional canned products. The major difference is the fact that the retortable pouch is a flexible container and therefore its shape is not defined which must be taken into account in the process design for pouched foods.

The most significant factor affecting the heating character of a pouched food is its thickness. Even slight increases in pouch thickness above that for which the process was designed may result in underprocessing. It is essential that methods be employed which will ensure that maximum thickness is not exceeded during retorting.

The best procedure for controlling pouch thickness is by physical confinement which places a maximum limit on thickness. Pouch confinement during processing is accomplished by specially designed retort racks, trays, or cassettes. For retail size pouches, thickness is typically maintained at ¾ in. Institutional pouches may have thicknesses of 1½–2 in.

Pouched foods can be adequately processed in an unconfined horizontal position if the thermal process is based on the maximum thickness attainable in commercial operation. These processes are usually longer than for pouches in the confined position and may result in lower product quality. For unconfined pouches, pouch thickness is largely determined by retort over-pressure, residual air in the pouch, fill weight, pouch orientation, and the product's physical characteristic (i.e., particulates, solid mass).

In addition to pouch thickness, the following critical factors must be considered in establishing a thermal process for pouched foods whether held in a confined or unconfined position.

1. *Residual air.* The air content of the pouch may not only affect thickness of unconfined pouches but also reduce the rate of the heat transfer from the heating medium to the product surface. This situation may occur if air is allowed to form a layer of insulation between the product surface and the interior wall of the pouch. The insulating effect of residual air is most significant when pouches are processed in a horizontal position.
2. *Pouch orientation.* Pouches confined to the same maximum thickness may show significant differences in heating as a result of pouch orientation (horizontal position).
3. *Product characteristics.* The heating character of the product, whether packaged in pouches, cans, or jars, will be affected by product formulation, viscosity, solid/liquid ratio and style of pack.

Process application: The retortable pouch is able to withstand only a minimal amount of internal pressure without seam failure. For this reason, the retort must be capable of providing overpressure during thermal processing and cooling.

Overpressure is critical to both seal integrity and pouch thickness when processing pouched foods in a horizontal unconfined position. In this instance, insufficient overpressure may result in under processing for two reasons. First, pouch expansion could increase product thickness and thereby reduce the heating rate, and secondly, the expanded pouches may adversely affect heat distribution in the retort by interfering with circulation of the heating medium.

Retort racks, tray, and carriers are used to confine or support individual pouches. These devices must be designed to provide equal exposure of each pouch to the heating medium during the process. Furthermore, the racks, in conjunction with the design features of the retort, must permit free circulation of the heating medium for maintenance of uniform temperatures throughout the retort.

In order to ensure adequate heat distribution, retort racks should provide channels between rows or layers of pouches for circulation of the heating medium. Heating rates will vary with the degree of circulation. Racks are construction of perforated material for optimum exposure.

Thermocouple placement. A cold spot determination is required to locate the slowest heating area for some products, taking into account all critical factors.

GENERAL METHOD (Bigelow et al. [5]). Measures the exact sterilizing value of a process, when such conditions as come-up-time, cooling water temperature, or the holding time after processing but before water cooling are different from normal retorting procedures. This method is also adapted to conditions when the heat penetration curve cannot be represented by one or two straight lines within the lethal temperature range on semi-logarithmic paper. It is not readily adapted to the calculation of processes when the retort temperature and/or initial temperature are different from those under which the heating data were obtained. Time and temperature data during the cooling cycle as well as the heating cycle must be recorded in order to use the graphical method.

BALL METHOD (Ball and Olson [31]). Is used when the heat penetration curve can be represented by not more than two straight lines on semi-logarithmic paper. The formula permits evaluating processes for retort and initial temperature conditions differing from those under which the heating data were obtained.

- Simple Curve
  - Straight line
- Broken Curve
  - Change from convection to conduction (or) Temperature gradients become so similar you cannot force water to flow past product; therefore, becomes conduction.

10.5 Thermal Processing of Retort Pouch: Summary

10.5.1 Critical Factors in the Thermal Processing of Retort Pouches

(a) Headspace air
  1. Product influence
  2. Air/pouch interaction
  3. Effect of air on process times

(b) Overriding air pressure
  1. Effects on pouch
  2. Operating parameters

(c) Pouch thickness
  1. Fill control
  2. Pouch confinement
  3. Influence on process time

(d) Pouch rack system
  1. Design
  2. Confinement method

(e) Other considerations
  1. Medium flow rate
  2. Initial temperature
  3. Pre-process factors
  4. Post-process factors

(f) Heat penetration tests
  1. Test design
  2. Thermocouple placement

(g) Temperature Distribution

(h) Processing System (Retort)

10.6 Food Regulations

10.6.1 United States Government Regulations

- Emergency Permit Control [21 CFR, Part 108]
- Good Manufacturing Practices [21 CFR, Part 110]
- Thermally Processed Low- Acid Foods [21 CFR, Part 113]
- Acidified Foods [21CFR, Part 114]
- USDA-Food Safety Inspection Services
  - Meat & Poultry Products [9 CFR, Part 318.300]

10.6.2 European Regulations

European Food Safety Authority (EFSA) is the keystone of European Union (EU) risk assessment regarding food and feed safety. In close collaboration with national authorities and in open consultation with its stakeholders, EFSA provides independent scientific advice and clear communication on existing and emerging risks.

The Food Standards Agency is an independent government department set up by an Act of Parliament in 2000 to protect the public's health and consumer interests in relation to food.

Food Safety Act 1990 Code of Practice

The General Principles of Food Law (Articles 5–10) entered into force on 21 February 2002 and must be followed when measures are taken. Existing food law principles and procedures must be adapted by 1 January 2007 in order to comply with the general framework established by Regulation EC/178/2002.

References

1. Hefnawy MM (2007) Developing Retorted Foods for Halal & Kosher Markets. Presented at Retort Pouch Europe 2007, The Packaging Group, 4-6 December 2007. Düsseldorf, Germany
2. Hefnawy MM (2007) Retort pouch foods, The next wave keynote speaker. Presented at retort pouch Europe 2007. The Packaging Group, Düsseldorf, 4–6 Dec 2007
3. Hu KJ, Nelson AI, Legault RR, Steinberg MP (1955) Feasibility of using plastic film packages for heat-processed foods. Food Technol 19(9):236–240
4. Nelson AI, Hu KH, Steinberg MP (1956) Heat process-able food films. Mod Packag 20(10):173–179
5. Keller RG (1959) Flexible packages for processed foods. Mod Packag 33(1): 145–149, 235
6. Tsutsumi Y (1974) The growth of food packed in retortable pouches in Japan. Presented at annual packaging institute forum, Chicago, 7 Oct 1974
7. Duxbury DD, Sams PF, Howeler WF, Gee JH, Miller WN (1970) Reliability of flexible packaging of flexible packaging for thermoprocessed foods under production conditions. Phase I: Feasibility. Technical Report 72-77-GP. U.S. Army, Natick
8. Omega Research Associates (1993) A critical analysis of the retort pouch 1982–1992, Justin

9. Hefnawy MM (2005) Thermal processing foods in retort pouch. Presented at second annual conference on retort pouch, Princeton, 17–19 May 2005
10. Hefnawy MM (2006) Processing/packaging foods in retort pouch. Presented at third annual conference on retort pouch, Atlantic City, 24–26 Apr 2006
11. Dunnigan J (2003) Retort pouches and combat superiority. StrategyWorld.com
12. Hefnawy MM (2006) Retort pouch foods in next ten years. Presented at third annual conference on retort pouch, Atlantic City, 18–20 Apr 2006
13. Lampi RA, Rubinate FJ (1973) Thermoprocessed foods in flexible packages: transition to production. Package Dev. 3(4):12–18
14. Hefnawy MM (2005) Formulating/processing foods in retort pouch. Presented at conference on food product development. Packaging Group, New Brunswick, 27 Oct 2005
15. Hefnawy MM (2007) What is retort pouch. Presented at retort pouch Europe 2007. The Packaging Group, Düsseldorf, 4–6 Dec 2007
16. Rubinate FJ (1973) Flexible packaging for heat-processed foods. Presented at symposium on flexible packaging for heat processed foods. The Packaging Institute, St. Louis, 27–29 Nov 1973
17. Hefnawy MM (1990) New product opportunities with thermal processed foods. Presented at international conference on new products, Thayngen, 4–5 Dec 1990
18. Long FE (1962) Flexible packages now withstand heat-processing temperatures of foods. Package Eng 7(3):63–80
19. Toyo Jidoki Co/Packaging Technologies and Inspection, LLC. Tuckahoe, New York. http://www.ptipacktech.com/products/toyojidoki/toyo_jidoki.asp
20. Herrmann Ultrasonics Inc., Bartlett. http://herrmannultrasonics.com/526.html
21. Multivac, Inc., Kansas City. http://www.multivac.com/our-products/thermoformers.html
22. Cryovac Food Packaging Co., Duncan. http://www.cryovac.com/NA/EN/food-packaging-equipment-vertical-form-fill.aspx
23. Stock America, Inc., Raleigh. http://www.stockamerica.com/retorts.asp, http://www.stockamerica.com/racks.asp
24. Long FE (1971) Automatic retort rack system for flexible packages. U.S. Patent 3,622,356
25. Allpax Products, LLC, Covington. http://www.allpax.com/products/spray-retorts/
26. Jackson JM, Shinn BM (1979) Fundamentals of food canning technology. AVI, Westport
27. Lampi RA (1979) Flexible packaging for thermoprocessed foods. In: Jackson JM, Shinn BM (eds) Fundamentals of food canning technology. AVI, Westport
28. Bigelow WD, Bohart GS, Richardson AC, Ball CO (1920) Heat penetration in processing canned foods. National Canners Association, Bulletin 16L
29. TechniCal Company, Kenner. http://www.tcal.com/CALSoftSoftware.php, http://www.tcal.com/CALPlexDataLogger.php
30. Mesa Laboratories, Inc., Lakewood. http://www.mesalabs.com/products-services/datatrace.html
31. Ball CO, Olson FCW (1957) Sterilization in food technology. McGraw-Hill, New York

Chapter 11
Inactivation Mechanisms of Different Mycotoxins

Faruk Bozoğlu

Abstract Mycotoxins can be inhibited at three strategic phases. During synthesis: by inhibiting the growth of producer strains and inhibiting the synthesis of mycotoxins. During processing: by physical separation, chemical inactivation and biological inactivation . During consumption: by alteration of bioavailability and modification of toxicity. Unquestionably, prevention of mycotoxin production is the best method for controlling mycotoxin contamination. Should the contamination occur, the hazard associated with the toxin must be removed if the product is to be used for food or feed purposes. Mycotoxins often occur in crops in the field prior to harvest. Physical separation on the field is the best choice to decrease the level of contamination. Enzymatic inactivation of fungal toxins is a beneficial strategy for the decontamination of agricultural commodities and for the protection of crops from phytotoxic effects of fungal metabolites. Natural compounds like ajowan, garlic and turmeric all have decreasing activity of mycotoxins. Microbial metabolites like aflastatin A (AsA), produced by A.parasiticus a novel inhibitor of aflatoxin production on melanin biosynthesis of Colletotrichum lagenarium was reported. Some toxin-producing fungi are able to degrade or transform their own products to non toxic forms under suitable conditions. Studies suggest that certain fungi, including A. parasitic us, degrade aflatoxins, possibly through fungal peroxidases. Fermentation with yeasts has also been effective in destroying patulin and rubratoxin B. Many of the studies have involved Lactobacillus strains, and physical binding has been proposed as one mechanism of mutagen removal. Sulfhydryl (thiol) compounds such as cysteine, N-acetylcysteine, reduced glutathione, and mercaptopropionylglycine interact with disulfide bonds of plant protease inhibitors and lectins via: sulfhydryl-disulfide interchange and oxidation-reduction reactions. Studies have shown that the major reaction product

F. Bozoğlu (✉)
Department of Food Engineering, Middle East Technical University, Ankara 06531, Turkey
e-mail: bozoglu@metu.edu.tr

M. Hefnawy (ed.), *Advances in Food Protection*, NATO Science for Peace and Security Series A: Chemistry and Biology 1, DOI 10.1007/978-94-007-1100-6_11,

formed from the reaction of aflatoxin B_1 with ammonium hydroxide at elevated temperature and pressure has lacked the lactone group characteristics of aflatoxin B1 which has no toxic activity.

Keywords Food safety • Mycotoxins • Fungal toxins

11.1 Introduction

Mycotoxins are toxic mold metabolites produced by toxigenic strains of different mold species. They have an important role in the occurrence of some human diseases such as liver cancer, chronic hepatitis and cirrhosis. When animals eat food-stuffs containing aflatoxin B1, these toxins will be metabolized and excreted as aflatoxin M1 in their milk. Aflatoxin M1 is resistant to thermal inactivation and is not destroyed completely by pasteurization, autoclaving or the other food processing procedures. Control of mycotoxins is the need of the hour, since their occurrence in foods and feeds is continuously posing threats to both health and economics all over the world. Since aflatoxin contamination is unavoidable, numerous strategies for their detoxification have been proposed. These include physical methods of separation, thermal inactivation, solvent extraction, adsorption from solution, microbial inactivation, and fermentation.

Chemical methods of detoxification are practiced as a major strategy for effective detoxification. Besides the post-harvest preventive measures, suitable detoxification methods are developed for inactivating or removing mycotoxins from the contaminated commodities, as the toxins are also produced by Aspergillus flavus and A. parasiticus even during post-harvest stages of crop production. Understanding of mechanisms of mycotoxin detoxification by physical, chemical and microbiological methods will enable establishment of combined treatment procedures to effectively decontaminate, contaminated foods and feeds. Such treatment methods are expected to be cost effective and minimally deleterious to food constituents.

Several physical and chemical detoxification methods developed so far have been critically discussed in different reviews for their advantages and limitations based on certain adopted strategies and specific criteria. Detoxification by microbiological means is also reviewed with a view to knowing the status on potential microorganisms and their enzymes that can degrade aflatoxins to less toxic or innocuous end products. Understanding of mechanisms of aflatoxin detoxification by physical, chemical and microbiological methods will enable establishment of combined treatment procedures to effectively decontaminate, contaminated foods and feeds. Such treatment methods are expected to be cost effective and minimally deleterious to food constituents.

11.2 Application of Ammonia

Ammoniation of corn, peanuts, cottonseed, and meals to alter the toxic and carcinogenic effects of mycotoxin contamination has been the subject of intense research effort by scientists in various government agencies and universities worldwide. Engineers have devised workable systems of treatment of whole seeds, kernels, or meals; chemists have identified and characterized products formed from the reaction of aflatoxin B1 with ammonia with and without a meal matrix; biochemists have studied the biological effects of these compounds in model systems; and nutritionists have studied animal responses to rations containing ammoniated or non-ammoniated components. The results of aflatoxin/ammonia decontamination research demonstrate the efficiency and safety of ammoniation as a practical solution to aflatoxin detoxification in foods and animal feeds [1].

Corn throughout the world is frequently contaminated by the fungus Fusarium moniliforme, which produces toxic fumonisins. Ammonia has been shown to detoxifying effect in corn and cottonseed. Since corn can be contaminated by both fumonisins and aflatoxins, application of ammoniation of corn either cultured with or naturally contaminated by F. moniliforme showed that fumonisin B1 levels in the culture material and in naturally contaminated corn were reduced by 30 and about 45%, respectively, by the treatment. Despite the apparent reduction in fumonisin content, the toxicity of the culture material in rats was not altered by ammoniation. Reduced weight gains, elevated serum enzyme levels and histopathological lesions, typical of F. moniliforme toxicity, occurred in rats fed either the ammoniated or non-ammoniated culture material. Atmospheric ammoniation of corn does not appear to be an effective method for the detoxification of F. moniliforme contaminated corn [2].

Although there was no significant change in dietary intake, body weight gain, and feed conversion ratio in chickens fed ammonia treated aflatoxin contaminated maize, these parameters were suppressed in birds fed aflatoxin-containing diet. These data suggest that replacement of aflatoxin-containing maize with ammoniated grains can significantly suppress aflatoxicosis, leading to improvement in production parameters in broilers [3, 4].

Rice, a cereal widely used for human and animal nutrition, is susceptible to aflatoxin contamination in the field and during storage. Therefore, the goal of the research was the evaluation of the efficacy and permanence of the ammoniation process through high pressure/high temperature (HP/HT) and atmospheric pressure/moderate temperature (AP/MT) conditions applied to rice samples artificially contaminated with aflatoxin B-1. For this purpose a 2(k) design was drawn up considering the temperature, the rice moisture and the process time as variables. Under both sets of conditions, aflatoxin B-1 concentration was reduced in a range of 90–100%. In conclusion, the process efficacy and permanence were achieved through the use of high temperature and long process time for both sets of conditions (HP/HT and AP/MT), respectively [5].

11.3 Chlorine Dioxide

The efficacy of chlorine dioxide (ClO_2) in detoxifying of the trichothecene mycotoxins verrucarin A and roridin A was evaluated. In the first experiment, verrucarin A (1, 5, or 10 mu/g) and roridin A (5 or 10 mu/g) were each inoculated onto square-inch sections of glass, paper, and cloth and exposed to 1,000 ppm of ClO_2 for either 24 or 72 h at room temperature. In the second experiment, verrucarin A and roridin A (1 or 2 ppm in water) were treated with 200, 500, or 1,000 ppm ClO_2 for up to 116 h at room temperature. Results for the first experiment showed that ClO_2 treatment had no detectable effect on either toxin.

For the second experiment, both toxins were completely inactivated at all tested concentrations in as little as 2 h after treatment with 1,000 ppm ClO_2. For verrucarin A, an effect was seen at the 500 ppm level, but this effect was not as strong as that observed at the 1,000 ppm level. Roridin A toxicity was decreased after treatment with 200 and 500 ppm ClO_2, but this was not significant until the 24-h exposure time was reached. These data show that ClO_2 in solution can be effective for detoxification of roridin A or verrucarin A at selected concentrations and exposure times [6].

11.4 Citric Acid

Chemical inactivation of aflatoxin B_1 (AFB_1) and aflatoxin B_2 (AFB_2) in maize grain by means of 1 N aqueous citric acid was confirmed by the AFLATEST™ immunoaffinity column method, high performance liquid chromatography (HPLC), and the Ames test (Salmonella-microsomal screening system). The AFLATEST™ assay showed that aflatoxins in the maize grain with an initial concentration of 29 ng/g were completely degraded and 96.7% degradation occurred in maize contaminated with 93 ng/g when treated with the aqueous citric acid. Aflatoxin fluorescence strength of acidified samples was much weaker than untreated samples as observed in HPLC chromatograms [7].

11.5 Sulfhydryl Compounds

Most food toxicants have specific groups responsible for their deleterious effects. Modifying such sites with specific a.acids, peptides and proteins lessens their toxicity. Sulfhydryl (thiol) compounds such as cysteine, N-acetylcysteine, reduced glutathione, and mercaptopropionylglycine interact with disulfide bonds of plant protease inhibitors and lectins via sulfhydryl-disulfide interchange and oxidation-reduction reactions. Such interactions with inhibitors from soybeans and with lectins from lima beans facilitate heat inactivation of the potentially toxic compounds, resulting in beneficial nutritional effects.

Related transformations of protease inhibitors in soy flour are also beneficial. Since thiols are potent nucleophiles, they have a strong affinity for unsaturated electrophilic centers of several dietary toxicants, including aflatoxins, sesquiterpene lactones such as elephantropin and parthenin, urethane, carbonyl compounds, quinones, and halogen compounds. Such interactions may be used in vitro to lower the toxic potential of the diet, and in vivo for prophylactic and therapeutic effects against oxidative damage. A number of examples are cited to illustrate the concepts and mechanisms of using sulfur amino acids to reduce the antinutritional and toxic manifestations of food ingredients [8].

11.6 Feed Additives

The possible protective effect of four feed additives against the toxic effects of T-2 toxin in growing broiler chickens was investigated in randomized trial consisting of six dietary treatments (control with no T-2 toxin or feed additive added, 2 ppm T-2 toxin alone, 2 ppm T-2 toxin plus 2.0 g/kg Mycofix, 2 ppm T-2 toxin plus 2.0 g/kg Mycosorb, 2 ppm T-2 toxin plus 2.5 g/kg MycoAd, and 2 ppm T-2 toxin plus 3.0 g/kg Zeolex). When no feed additive was included, 2 ppm dietary T-2 toxin significantly decreased BW and increased feed:gain ratio. When 2.0 g/kg Mycofix were added to the diet, the feed additive protected against the adverse effects of T-2 toxin on BW, BW gain, and feed:gain ratio; however, no protection against the adverse effects of T-2 toxin on final BW and BW gain were obtained by supplementation of any of the other three feed additives. The results of trial indicate that the only feed additive capable of counteracting the adverse effects on performance caused by the dietary administration of 2 ppm T-2 toxin was the additive based on the enzymatic inactivation of the 12,13-epoxide ring of the trichothecenes. This study also confirms previous reports showing that aluminosilicates are not effective against trichothecene mycotoxins [9].

Aqueous extract of ajowan seeds was found to contain an aflatoxin inactivation factor (IF). An approximate 80% reduction in total aflatoxin content over the controls was observed. This observed phenomenon of reduction in total toxin was referred to as toxin inactivation. Temperature was found to influence the rate of toxin inactivation. At 45°C, it was found to be rapid during the initial 5 h and slowed later. The IF was found to retain considerable activity even after boiling and autoclaving, indicating partial heat stability. Toxin decontamination in spiked corn samples could be achieved using IF. This study emphasizes the potential of ajowan IF in aflatoxin removal from contaminated food commodities. However, the biological toxicity, if any, of the IF inactivated aflatoxins needs to be confirmed [10].

11.7 Biological Detoxification

Some toxin-producing fungi are able to degrade or transform their own products under suitable conditions. Pure cultures of bacteria and fungi which detoxify mycotoxins have been isolated from complex microbial populations by screening and enrichment culture techniques. Genes responsible for some of the detoxification activities have been cloned and expressed in heterologous hosts. The detoxification of aflatoxins, cercosporin, fumonisins, fusaric acid, ochratoxin A, oxalic acid, patulin, trichothecenes and zearalenone by pure cultures were also reported [11].

11.8 Extrusion Process

Cottonseed is an economical source of protein and is commonly used in balancing livestock rations; however, its use is limited by protein, fat, gossypol, and aflatoxin contents. The extrusion temperature study showed that aflatoxin levels were reduced by an additional 33% when the cottonseed was extruded at 160°C as compared to 104°C. Furthermore, the multiple-pass extrusion study indicated that aflatoxin levels were reduced by an additional 55% when the cottonseed was extruded four times as compared to one time. Total estimated reductions of 55% (three stages of processing at 104°C), 50% (two stages of processing at 132°C), and 47% (one stage of processing at 160°C) were obtained from the combined equations. If the extreme conditions (four stages of processing at 160°C) of the evaluation studies are applied to the combined temperature and processing equation, the resulting aflatoxin reduction would be 76% [12].

Traditional nixtamalization and an extrusion method for making the dough (masa) for corn tortillas that requires using lime and hydrogen peroxide were evaluated for the detoxification of aflatoxins. The traditional nixtamalization process reduced levels of aflatoxin B-1 (AFB(1)) by 94%, aflatoxin M-1 (AFM(1)) by 90% and aflatoxin B-1- 8,9-dihydrodiol (AFB(1)-dihydrodiol) by 93%. The extrusion process however reduced levels of AFB(1) by 46%, AFM(1) by 20% and AFB(1)-dihydrodiol by 53%. Extrusion treatments with 0, 0.3 and 0.5% lime reduced AFB(1) levels by 46, 74 and 85%, respectively.

The inactivation of AFB(1), AFM(1) and AFB(1)- dihydrodiol in the extrusion process using lime together with hydrogen peroxide showed higher elimination of AFB(1) than treatments with lime or hydrogen peroxide alone. The extrusion process with 0.3% lime and 1.5% hydrogen peroxide was the most effective process to detoxify aflatoxins in corn tortillas, but a high level of those reagents negatively affected the taste and aroma of the corn tortilla as compared with tortillas elaborated by the traditional nixtamalization process [13].

Samples of corn flour experimentally contaminated with AFB1 (50 ppb) and deoxynivalenol (DON) (5 ppm) were extruded. The effects of three extrusion variables (flour moisture, extrusion temperature and sodium metabisulphite addition) were analysed according to a two-level factorial design. The process was effective for the reduction of DON content (higher than 95%) under all the conditions assessed, but was only partially successful (10–25%) for the decontamination of AFB1.

The results show that extrusion cooking is effective for the inactivation of DON but is of limited value for AFB1, even if metabisulphite is added. More severe extrusion conditions are needed for the detoxification of AFB1. As contamination with DON occurs mainly in the field prior to harvesting and that of AFB1 is normally produced during grain storage, maize is often contaminated with DON but not with AFB1. Under these conditions, the described extrusion process can be used for the detoxification of DON. The addition of sodium metabisulphite did not significantly affect the inactivation of AFB1. Extrusion cooking is, therefore, an appropriate treatment for DON-contaminated maize in countries; because of the prevailing conditions, these are the only toxins present [14, 15].

11.9 Miscellaneous

Aflatoxins are also sensitive to UV light. Exposure of artificially contaminated milk to UV light inactivated 3.6–100% of AFM1 in the milk depending on the exposure time and in dried figs artificially contaminated with AFB1 reduced the toxin level by 45.7%.

Solar energy is also widely used and shown to decrease the amount of aflatoxins from 30–80% in peanut cakes and flakes peanut oil and olive oils indifferent parts of the world . High hydrostatic pressure application is another method to inactivate mycotoxins present in foods, however pressure exciting 500 MPa has detrimental effects on the food itself.

11.10 Conclusive Remarks

Although mycotoxins can be inactivated by different methods the application of these methods depends on the type of mycotoxin or food/feeds to be consumed. Inactivation of these toxic compounds are mainly due to changes occurring in their chemical structures by the application of inactivation methods.

While pressure and heat degrades molecules to smaller structures, chemical applications opens the ring structures resulting inactivated compounds. Treatment by special additives have no effect on the structures but preferred due to absorption of toxic materials that results decreasing bioavailability of these toxic compounds.

References

1. Park DL (1993) Perspectives on mycotoxin decontamination procedures. Food Addıt Contam 10(1):49–60
2. Norred WP, Bacon CW, Plattner RD, Vesonder RF (1991) Differential cytotoxicity and mycotoxin content among isolates of Fusarium-moniliforme. Mycopathologia 115(1):37–43
3. Allameh A, Safamehr A, Mirhadi SA, Shivazad M, Razzaghi-Abyaneh M, Afshar-Naderi A (2005) Evaluation of biochemical and production parameters of broiler chicks fed ammonia treated aflatoxin contaminated maize grains. Anim Feed Scı Technol 122(3–4):289–301

4. Norred WP, Voss KA, BaconCW RRT (1991) Effectiveness of ammonia treatment in detoxification of Fumonisin contaminated corn. Food Chem Toxıcol 29(12):815–819
5. Trujillo FRM, Yepez AJM (2003) Efficacy and stability of ammoniation process as aflatoxin B-1 decontamination technology in rice (Oriza sativa). Arch Latinoam Nutr 53(3):287–292
6. Wilson SC, Brasel TL, Martin JM, Wu C, Andriychuk L, Douglas DR, Cobos L, Straus DC (2005) Efficacy of chlorine dioxide as a gas and in solution in the inactivation of two trichothecene mycotoxins. Int J Toxıcol 24(3):181–186
7. Mendez-Albores A, Arambula-Villa G, Loarea-Pina MGF, Castano-Tostado E, Moreno-Martinez E (2005) Safety and efficacy evaluation of aqueous citric acid to degrade B-aflatoxins in maize. Food Chem Toxıcol 43(2):233–238
8. Friedman M (1994) Mechanism of beneficial-effects of sulfur amino-acids sulfur compounds in foods. Acc Sym Ser 564:258–277
9. Diaz GJ, Cortes A, Roldan L (2005) Evaluation of the efficacy of four feed additives against the adverse effects of T-2 toxin in growing broiler chickens. J Appl Poultry Res 14(2):226–231
10. Hajare SS, Hajare SN, Sharma A (2005) Aflatoxin inactivation using aqueous extract of ajowan (Trachyspermum ammi) seeds. J Food Scı 70(1):29–34
11. Karlovsky P (1999) Biological detoxification of fungal toxins and its use in plant breeding, feed and food production. Nat Toxins 7(1):1–23
12. Buser MD, Abbas HK (2002) Effects of extrusion temperature and dwell time on aflatoxin levels in cottonseed. J Agric Food Chem 50(9):2556–2559
13. Elias-Orozco R, Castellanos-Nava A, Gaytan-Martinez M, Figueroa-Cardenas JD, Loarca-Pina G (2002) Comparison of nixtamalization and extrusion processes for a reduction in aflatoxin content. Food Addit Contam 19(9):878–885
14. Cazzaniga D, Basilico JC, Gonzalez RJ, Torres RL, de Greef DM (2001) Mycotoxins inactivation by extrusion cooking of corn flour. Lett Appl Mıcrobıol 33(2):144–147
15. Basappa SC, Shantha T (1996) Methods for detoxification of aflatoxins in foods and feeds – a critical appraisal. J Food Scı Technol 33(2):95–107

Chapter 12
Pulsed Electric Fields to Obtain Safe and Healthy Shelf-Stable Liquid Foods

Ingrid Aguiló-Aguayo, Robert Soliva-Fortuny, Pedro Elez-Martínez, and Olga Martín-Belloso

Abstract Pulsed electric fields (PEF) technology provides the potential of ensuring safety and maintaining the physico-chemical quality of liquid food products without substantially impacting the content and composition of thermolabile compounds. This is especially relevant in the case of plant-based foods, because some of the features that are currently most appreciated by consumers, such as aroma or bioactive potential, are related to this heat-sensible fraction. Specifically, fruit juices and vegetable-based beverages exhibit a remarkable content in phytochemicals with health-promoting benefits, some of them with a significant antioxidant potential. Although the effectiveness of PEF treatments has been extensively studied during the past couple of decades, their impact on the bioactive composition of foods is still being researched.

Through the presentation, some of the key factors that rule the inactivation/destruction of health-related constituents in foods will be introduced and discussed. Recently published research results will be reviewed and compared with those obtained for other thermal and non-thermal processing technologies, with a special stress on the effect of PEF-processing variables on the bioactive composition of foods throughout their whole shelf-life. Furthermore, different examples will be presented to illustrate not only the potential but also the limitations of PEF technology when aiming at preserving the health-promoting features of plant-based foods.

Keywords Food safety • Pulsed electric fields • Shelf-stable liquid foods

---

I. Aguiló-Aguayo, R. Soliva-Fortuny, P. Elez-Martínez, and O. Martín-Belloso (✉)
Department of Food Technology, University of Lleida, Avda. Alcalde Rovira Roure 191, Lleida 25198, Spain
e-mail: rsoliva@tecal.udl.cat; omartin@tecal.udl.cat

M. Hefnawy (ed.), *Advances in Food Protection*, NATO Science for Peace and Security Series A: Chemistry and Biology 1, DOI 10.1007/978-94-007-1100-6_12,

12.1 Introduction to Pulsed Electric Fields Processing Basics

Non-thermal processes have gained importance in recent years due to the increasing demand for foods with a high nutritional value and fresh-like characteristics, representing an alternative to conventional thermal treatments. Pulsed electric fields (PEF) is an emerging technology that has been extensively studied for non-thermal food processing.

12.1.1 Historical Background

First applications of electrical current for food treatment took place at the end of the nineteenth century when Prochownick and Spaeth [1] investigated the bactericidal effects of direct and alternating electrical current. In the 1920s a process called 'Electropure' was introduced in Europe and the USA for milk pasteurization in order to improve consumer health [2–4]. The process involved heating the milk to a temperature of 70°C, and then passing it through carbon electrodes in an electric heating chamber to inactivate *Mycobacterium tuberculosis* and *Escherichia coli*. Pulsed discharges of high voltage electricity across two electrodes for microbial inactivation were first investigated in the 1950s through the so-called 'electrohydraulic treatment' [5, 6]. The electrodes were submerged in the liquid medium within a pressure vessel and electric arcs were generated causing a bactericidal effect but leading to food contamination by the disintegration of food particles and electrodes. In 1960, Gossling reported microbial inactivation for *Streptococcus lactis* dependent on treatment intensity and proposed small scale batch and continuous treatment chambers [7]. Doevenspeck [8, 9] described the application of pulsed electric fields (PEF) for disruption of cells in food material to improve phase separation. It was found that the application of electric fields of low intensity (pulses below 200 V/mm) led to enhanced growth of *E. coli*, whereas increasing electric field strength resulted in cell death. The first systematic studies on the non-thermal lethal effect of homogeneous pulsed electric fields on microbes were conducted by Sale and Hamilton [10]. They observed that electric field strength and total treatment time were the most important factors involved in the inactivation of bacteria such as *E. coli*, *Staphylococcus aureus*, *Micrococcus lysodeikticus*, *Sarcina lutea*, *Bacillus subtilis*, *Bacillus cereus*, *Bacillus megaterium*, and *Clostridium perfringens*, and yeasts such as *Saccharomyces cerevisiae* and *Candida utilis*. Damage to the cell membrane, causing an irreversible loss of its function as a semipermeable barrier between the cell and its environment, was proposed as the cause of the cell death. The fundamental effects and mechanisms of electric fields on the disruptive effect on biological cells were introduced by Zimmermann et al. [11] and Neumann and Rosenheck [12]. The cell membrane was considered to act as a capacitor containing a perfectly elastic dielectric. The electro-mechanical instability observed due to the influence of an external electric field on biological cells is still one of the most accepted theories for microbial inactivation. Tsong [13] also suggested that pore formation might occur as a consequence of structure defects within the cell membrane, expanding spontaneously formed pores in the presence of an electric field. The research group lead by Prof. Hülsheger developed mathematical models to describe microbial inactivation by PEF as a function of electric field strength and treatment time [14, 15].

In 1986, a system named ELSTERIL®, consisting of a high voltage pulse generator with a peak voltage of 15 kV and a repetition rate of 22 Hz, was developed for the electric pasteurization of liquid products. The storage capacity ranged between 0.5 and 5 μF, and an ignitron switch was used to discharge the stored electrical energy [16]. Dunn and Pearlman [17] designed a circular parallel-plate chamber for both static and continuous PEF treatments with stainless steel electrodes and a nylon spacer. Reductions of more than five log cycles were reported in the microbial populations of naturally occurring microorganisms in orange juice after applying treatments of 33.6–35.7 kV/cm and a process temperature of 42–65°C. Moreover, different geometric configurations of electrodes such as parallel plate, needle-plate, wire-cylinder and rod-rod shaped electrodes were studied for the destruction of living cells [18]. Coaxial and cofield continuous flow treatment chambers were later designed [19, 20] and are still currently used due to their simplicity in structure.

In 1995, Pure Pulse Technologies (San Diego, USA) registered a patent describing a continuous processing system working in the range of 10–25 kV/cm for 1–100 μs for the treatment of pre-heated liquid foods [17, 21]. In the same year, the Washington State University (Pullman, WA, USA) registered a patent of a continuous lab-scale PEF system [22]. Studies on raw peach juice, skim milk, beaten eggs, pea soup, apple juice and reconstituted apple juice were conducted with this system at electric field strengths of 25–45 kV/cm [23].

The first reviews on the application of PEF technology were published in the 1990s ([24–28]) and the first monographic book was published in 1999 [29]. Extensive research studies have been carried out for different food products regarding microbial inactivation [22, 29–31], engineering aspects of processing [22, 32, 33], effects on food matrices [34], inactivation of enzymes [35–38], effects on the bioactive potential of foods [38, 39] or induction of stress reactions and secondary metabolite production [39, 40]. Nowadays, more than 20 research groups are working actively in this area worldwide.

12.1.2 Fundamentals of PEF Processing

PEF processing involves the application of a high intensity electric field (20–80 kV/cm) in the form of short pulses (μs or ms) to a food placed between two electrodes. PEF technology provides fresh-like and safe foods while reducing quality losses that can be triggered after thermal processing [43, 44]. Application of continuous PEF processing is not suitable for solid food products that do not allow pumping and is restricted to low-conductive food products without air bubbles.

In addition, the maximum particle size in the product must be smaller than the gap of thetreatment region in the chamber in order to ensure proper treatment [45]. Liquid food products are susceptible to be treated by PEF because they are primarily composed by water and nutrients, which are electrical conductors due of the presence of large concentrations of salts and dipolar molecules. When subjected to a pulsed electric field, polarization of dipolar molecules and bulk movement of charge carriers, such as ions, induce a capacitive and a resistive current [46, 47]. This leads to dielectric breakdown of the microbial cell membranes and to interaction with the charged molecules of food [48, 49]. Hence, PEF technology has been suggested for the pasteurization of foods such as juices, milk, yogurt, soups, and liquid eggs [37, 50–53].

In general, a continuous PEF treatment system is composed of treatment chambers, a pulse generator, a fluid-handling system, and monitoring systems (Fig. 12.1) [54, 55]. The PEF treatment chamber houses the electrodes delivering high voltage discharges to the food material placed between them. Insulating material holds the electrodes in fixed positions and forms the chamber containing the food to be processed. The proper design of the treatment chamber is essential for the efficiency of PEF processing [56]. The pulse generator supplies high voltage pulses to foods flowing through the treatment chamber. After the treatment, the food is aseptically packaged and stored under refrigeration. Temperature- and pulse-monitoring systems are used to supervise the process. Temperature is controlled with thermocouples, while pulses are monitored with high-voltage probes, current monitors, and oscilloscopes [57].

The effectiveness of PEF technology not only depends on the type of equipment used but also on the treatment parameters and media to be processed [29]. Electric field strength, total treatment time, pulse shape, pulse frequency, pulse width, polarity and temperature are operational parameters reported to affect the efficiency of the process. Electric field strength and treatment time have been identified as the most relevant factors. The effectiveness of microbial and enzyme inactivation by PEF is higher when electric field strength and treatment time increases [53–55, 58–63]. PEF treatments are applied in the form of short pulses to avoid excessive heating or undesirable electrolytic reactions.

The electric field may be applied in the form of exponentially decaying, square wave, or oscillatory pulses [53]. Square-wave pulse appears to be the most energy-efficient and lethal wave pulse shape (Qin et al. 1995). However, according to Zhang et al. [22], pulse width must be determined within the range that does not ramp up the food temperature. If pulse width is too long, food temperature rises to an undesirable level for a PEF treatment to be considered non thermal. Pulse frequency also has a role in determining the efficiency of the treatment. Previous studies have shown the simultaneous influence of frequency and pulse width on the reduction of enzyme activity in different products [36, 37, 54, 65, 66]. Moreover, the use of pulses of alternating polarity reduces the risk of undesirable electrochemical reactions as well as formation of deposits at the electrodes, resulting in a more uniform PEF treatment [22]. Bipolar pulses are reported to be more efficient than monopolar pulses on the destruction of several microorganisms in fluid products

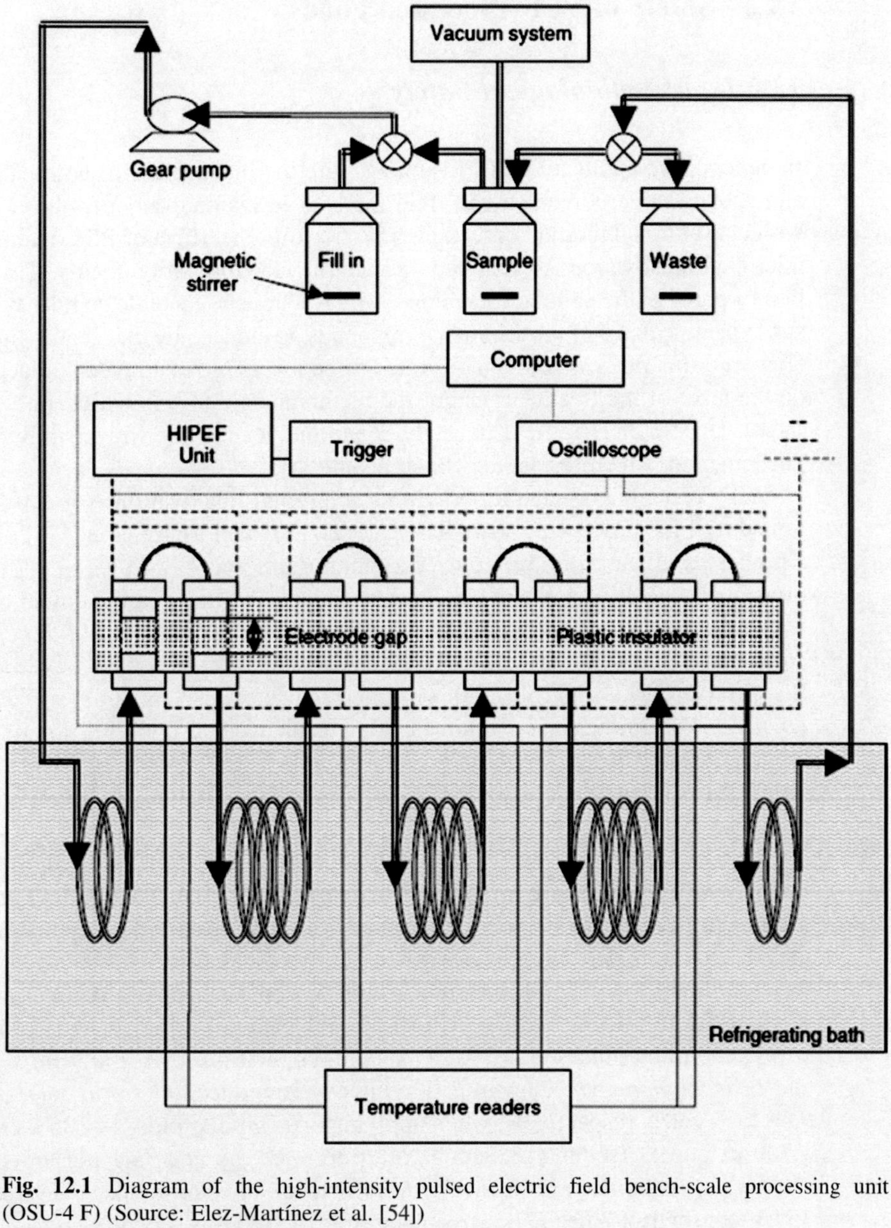

Fig. 12.1 Diagram of the high-intensity pulsed electric field bench-scale processing unit (OSU-4 F) (Source: Elez-Martínez et al. [54])

([19, 50], Ho et al. 1995). In contrast, controversial results regarding polarity have been obtained depending on the enzyme and the food under study. Pectin methylesterase (PME) inactivation in orange juice subjected to bipolar treatments was greater than when monopolar pulses were applied [65]. However, PME activity in tomato and strawberry juice was not affected when a PEF treatment was applied in mono- or bipolar mode [36, 61].

12.2 Safety of PEF-Processed Foods

12.2.1 Microbiological Safety

In general, the application of PEF may assure the inactivation of both pathogenic and spoilage microorganisms in fluid foods at levels equivalent to those attained with conventional thermal processing [52, 67–76]. The effect of PEF treatments on microbial inactivation is explained through the electroporation theory. The formation of pores in the cellular membrane, which is able to generate cellular lysis with subsequent leakage of intracellular compounds, is induced by the electric field [10]. However, this process has demonstrated to be reversible or irreversible depending on the level of the treatment and/or the membrane organizational changes (Tsong 1990). Drastic electric field intensity conditions lead to the irreversibility of this phenomenon, which results in cellular death [35].

Many researchers have reported that microbial inactivation is greatly conditioned by PEF processing variables. Iu et al. [77] and Liang et al. [78] obtained a higher inactivation of *Salmonella* Typhimurium and *E. coli* O157:H7 populations in orange juice and apple cider, respectively, by increasing number of et al. [72] reached higher microbial inactivation of *E. coli* O157:H7, *E. coli* and *Salmonella* Enteritidis in several fruit juices when treatment time was increased. Elez-Martínez et al. [50, 68] observed an increase in *S. cerevisiae* and *Lactobacillus brevis* inactivation in orange juice when applying treatments at low frequency. Moreover, bipolar pulses are generally reported to be slightly more efficient than monopolar pulses on the inactivation of microorganisms ([19, 68], Ho et al. 1995).

The optimization of PEF treatment conditions has contributed to the standardization of PEF processing for the destruction of microorganisms in fluid products such as milk or juices. Consistently, Sobrino-López et al. [75] obtained a maximal inactivation of 4.5 log cycles when 150 bipolar pulses of 8 μs at 35 kV/cm were applied to skim milk inoculated with S. aureus. Elez-Martínez et al. [50], reported up to 5.8 log reductions in *L. brevis* counts in orange juice processed with 250 bipolar pulses of 4 μs at 35 kV/cm. Populations of *S. Enteritidis*, *E. coli* and *Listeria monocytogenes* in melon juice were reduced by up to 3.71, 3.7 and 3.56 log cycles, respectively, when applying 4-μs bipolar pulses of 35 kV/cm for 1,709 μs at 193 Hz, and reductions of up to 3.56, 3.6 and 3.41 log units of the same microorganisms, respectively, were reached in watermelon juice treated at 35 kV/cm for 1,682 μs [72]. A treatment time of 1,000 μs was needed for 35-KV/cm pulses applied at 100 Hz to reduce *Salmonella* Enteritidis in tomato juice by 4.184 log cycles [74]. Inactivation of microorganisms by PEF can not only be affected by the processing conditions but also by the fluid medium, the target microorganism and the microbial characteristics [79]. Mosqueda-Melgar et al. [73] reported that *E. coli* O157:H7 in acidic juices exhibits high sensibility to PEF. Electrical conductivity and pH may also have an influence on microbial destruction. Inactivation rates of microorganisms such as *L. brevis*, *E. coli*,

S. cerevisiae, *Salmonella* Dublin and *Listeria innocua* increased with decreasing conductivity of the treatment medium [30, 80–82]. Juices usually have lower conductivity and pH than milk, thus allowing a greater inactivation of microorganisms. In low-conductive foods, differences in conductivity between the microorganisms cytoplasm and the medium causes an additional pressure on the cell membrane due to osmotic forces, which makes it more sensitive to the PEF treatment [81]. Moreover, it has been reported that low water activity values act as a protective factor against the inactivation of microorganisms due to PEF treatments [83, 84].

Several authors have demonstrated that Gram-negative bacteria are more susceptible to PEF inactivation than Gram-positive bacteria [85–89]. This fact could be related to the composition of their membranes, since the cell membranes of Gram-positive bacteria are more rigid and thicker than those of Gram-negative microorganisms. This could constitute an additional protection against the PEF treatment [83]. In general, bacterial spores are resistant to PEF treatments, but after germination they become PEF-sensitive [90, 91]. Moreover, the type of bacteria also seems to affect the effectiveness of the treatment. Evrendilek et al. [51] observed that PEF treatment conditions (22 kV/cm for 216 μs) with lethal effects against *Saccharomyces uvarum* and *Rhodotorula rubra* inoculated in beer were less severe than those required to satisfactorily inactivate a bacterial culture of *Lactobacillus plantarum*, *Pediococcus damnosus*, and *B. subtilis* (41 kV/cm for 175 μs). Microbial inactivation rates have been shown to be dependent on the initial microbial concentration and the growth stage. Cells in logarithmic phase were found to be more sensitive to PEF treatments than those in stationary phase [92]. In contrast, Zhang et al. [22, 93] reported that initial cell concentration of the microorganism was inversely correlated with its survival fraction after a PEF treatment.

The incidence of uncontrolled variables such as the increase in the temperature medium (attributed to the electric energy conversion into heat) would enhance the efficiency of PEF processing. Liang et al. [78] reported more than 5.0 log reductions of *Salmonella* Typhimurium in fresh squeezed orange juice after a PEF treatment of 90 kV/cm, a pulse number of 20 and an outlet temperature of 55°C. Likewise, McDonald et al. [94] reached up to 5.0 log reductions in the counts of E.coli in orange juice by applying a treatment of six 30-kV/cm pulses per volume with an outlet temperature of 54°C. The combined application of PEF and heat on milk has been shown to be synergistic [95]. Walkling-Ribeiro et al. [96] reduced the microbial counts in milk by up to 6.0 log cycles by combining 50 kV/cm for 33 μs and a treatment temperature of 50°C, thus extending the product shelf-life to 21 days at 4°C.

On the other hand, the combination of PEF treatments with bacteriocins or other antimicrobials opens innovative possibilities for application on low acidic products in a hurdle approach. The survival of *S. aureus* inoculated into skim milk and treated with a combination of nisin (a bacteriocin from lactic acid bacteria) and PEF has been evaluated. Roughly 6.0 log microbial reductions were reported when combining a PEF treatment of 35 kV/cm for 2,400 μs with 20 ppm nisin, whereas over 4.5 log reductions where attained when 35 kV/cm for 240 μs where applied in combination with 150 ppm nisin [97]. Furthermore, the use of natural antimicrobial substances such as organic acids and plant essential oils could enhance the antimicrobial effect of PEF treatments in fruit juices [77, 98]. Combinations of PEF (35 kV/cm for 1,000 μs at 100 Hz and 4 μs pulse length in bipolar mode) with 2.0% citric acid or 0.1% cinnamon bark oil were needed for inactivating *Salmonella* Enteritidis by more than 5.0 log units [99]. Populations of *E. coli* O157:H7, *Salmonella* Enteritidis and *L. monocytogenes* were reduced by more than 5.0 log units in PEF-processed melon juice (35 kV/cm for 1,709 μs at 193 Hz and 4 μs pulse duration) and watermelon juice (35 kV/cm for 1,682 μs at 193 Hz and 4 μs pulse duration) containing 2.0% and 1.5% citric acid, respectively, or 0.2% cinnamon bark oil. In addition, these treatments were also able to inactivate mesophilic, psychrophylic, and molds and yeasts populations, leading to a shelf-life or more than 91 days in both juices stored at 5°C [99].

12.2.2 Chemical Risks

Some authors have investigated if PEF processing could chemically induce changes in the quality of food or even cause harmful effects on the consumers. Hülsheger and Niemann [14] suggested that hypochloric acid, generated from chloride in buffers under the action of an electric field, could contribute to the lethality of PEF treatment. Nevertheless, Wouters et al. [100] later demonstrated that the inactivation degree of *L. monocytogenes* in buffer systems containing chloride ions was no significantly higher than in buffers without chloride. They argued that the length of the pulse is a critical parameter when aiming at controlling undesirable electrochemical reactions. Some studies have identified the unintentional emission of metal as a significant risk [32]. Charged electrodes are in contact with the food and, therefore, the formation of electrolytic products and the release of electrode material into the product stream can hardly be controlled. Thus, the particular design of electrodes and treatment devices, as well as pulse shape definition, is critical to prevent these risks [101]. The amount of metals released depends on the type of product, the specific composition of the electrodes and pulse geometry. On the other hand, changes in food components could be induced due to the electrochemical action. Detailed information on the composition of foods can be obtained by chemical fingerprinting [102]. Minor chemical changes have been demonstrated to be induced in PEF-treated tomato product after processing [103].

From a legal perspective, PEF treatment is subjected to the Novel Food Regulation (NFR) issued by the European Union [104]. The NFR states that safety of food products can be assumed if during the novel process no additives are introduced, and the resulting product does not substantially differ from the untreated product.

12.3 Effect of PEF-Treatments on the Health-Related Potential of Food

Studies evaluating the effects of PEF processing on bioactive components of food are more limited compared to those on microbial inactivation. Recent reports point out that PEF appear to have a light effect on the phenolic composition of foods. Sánchez-Moreno et al. [105] reported no changes in the total flavanone content of orange juice after applying a PEF treatment at 35 kV/cm for 450 μs with 4-μs bipolar pulses at 800 Hz. Anthocyanin retention in HIPEF-treated strawberry juice ranged from 96.1% to 100.5% when exposed to electric field strengths from 20 to 35 kV/cm for up to 2,000 μs [40]. Odriozola-Serrano et al. [106, 107] studied the effect of PEF processing (35 kV/cm for 1,500 μs with bipolar pulses of 4-μs at 100 Hz) and storage on individual phenolic compounds of tomato juice. No significant differences among PEF- and thermally-treated juices was observed just after processing (Fig. 12.2). However, PEF-processed tomato juices exhibited higher retention of their total phenolic content during 56 days than heat-treated juices (90°C for 60 s). In strawberry juice, PEF-treated samples (35 kV/cm for 1,700 μs with bipolar pulses of 4-μs at 100 Hz) maintained higher concentrations of phenolic acids (ellagic and p-coumaric acid) during the storage period than those thermally-treated [108].

Among vitamins, vitamin C has been extensively studied due to its heat-lability. In general, reports indicate that vitamin C is less affected by PEF treatments than by conventional treatments in very different products such as orange juice [57, 60, 105, 109], apple juice [70], strawberry juice [108], tomato juice [106, 110], orange-carrot juice [111], and fruit juices-soy milk blend [71]. On the other hand, no sig-

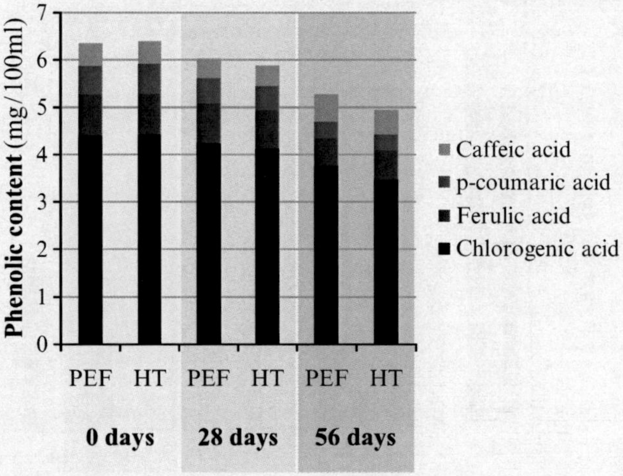

Fig. 12.2 Contents of the main hydroxycinnamic acids in PEF-treated (35 kV/cm for 1,500 μs) and heat-treated (HT: 90°C for 1 min) tomato juice stored under refrigeration (4°C) (Adapted from Odriozola-Serrano et al. [107])

nificant effect of PEF on both water-soluble vitamins (riboflavin and thiamine) and fat-soluble vitamins (cholecalciferol and tocopherol) from milk was reported by Bendicho et al. [112].

The effects of PEF processing on the antioxidant potential of various fruit juices have been evaluated through changes in the concentration of carotenoid compounds. Cortés et al. [113] observed that individual carotenoids such as β-carotene, β-cryptoxanthin, zeaxanthin and lutein from orange juice did not change after applying bipolar PEF treatments at 25–40 kV/cm for 30–340 μs. In addition, they reported that the decrease in the concentrations of total carotenoids during the cold storage of untreated and thermally-treated (90°C for 20 s) juices was greater than that of PEF-treated samples. This is consistent with the results reported by Quitão-Teixeira et al. [114] for PEF treated carrot juice (Fig. 12.3). On the other hand, Odriozola-Serrano et al. [107] observed an increase in the lycopene content of tomato juice after applying 35 kV/cm during 1,000 μs with bipolar pulses of 7 μs at 250 Hz. These results are in accordance with those reported by Oms-Oliu et al. [115], who showed that bipolar pulses applied at higher frequencies led to a lycopene content in treated watermelon juice higher than in the untreated juice. Torregrosa et al. [116] also reported that HIPEF treatments set up at 25 or 30 kV/cm caused a significant enhancement in the content of carotenoids in orange-carrot juice. In contrast, results obtained by Zulueta et al. [117] showed that PEF treatments at 15 kV/cm caused a slight increase in the concentrations of some carotenoids in an orange juice-milk beverage, while slight decreases occurred when treating at electric field strengths of 40 kV/cm.

Odriozola-Serrano et al. [107] observed the maximum values of antioxidant capacity in tomato juice when applying bipolar pulses at frequencies higher than

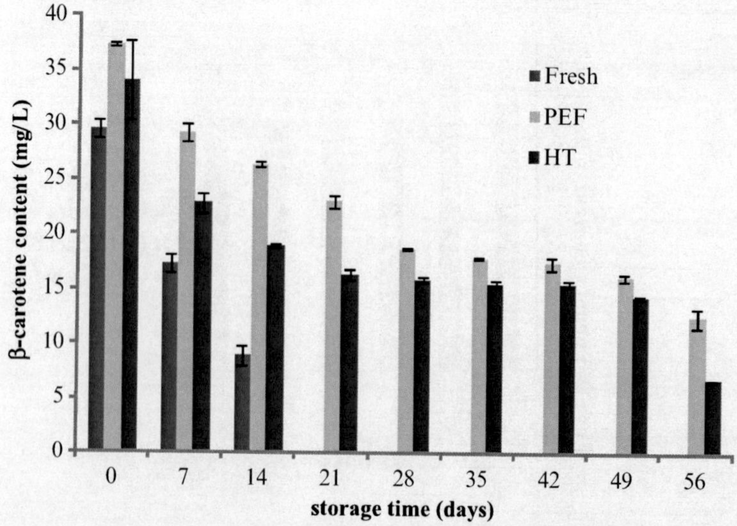

Fig. 12.3 Effects of PEF and heat (HT) pasteurizing treatments on β-carotene retention of carrot juice throughout storage at 4°C for 56 days (Adapted from Quitão-Teixeira et al. [114])

150 Hz and pulse widths lower than 5 μs. Elez-Martínez and Martín-Belloso [118] reported that HIPEF treatments of 15 or 35 kV/cm for 1,000 μs at 200 Hz applying bipolar of monopolar 4-μs pulses led to similar levels of antioxidant capacity in orange juice and 'gazpacho' soup compared with a conventional thermal treatment. Sánchez-Moreno et al. [109] observed that the antioxidant capacities of untreated and PEF-treated orange juices were not significantly different. Some differences among studies could be related to the different equipment and raw materials used as well as to the differences in the treatment conditions.

The effect of PEF treatments on food proteins has been scarcely studied. Martín-Belloso et al. [119] reported no protein coagulation when liquid whole egg was exposed to up to 100 square-wave pulses of 2–4 μs using an electric field intensity of 26 kV/cm. Li et al. [120] did not observe changes in the secondary structure of milk proteins and loss of immunoactivity after applying bipolar treatments of 41.1 kV/cm for 54 μs or heat treatments. Odriozola-Serrano et al. [39] reported that whey proteins (serum albumin, β-lactoglobulin and α-lactalbumin) were similarly retained in PEF and heat-treated milk. In the same way, it has been observed that PEF induce little changes in the fatty acids profile. Garde-Cerdán et al. [121] observed that PEF treatments of 35 kV/cm for 1 ms rendered grape juices with a content of total fatty acids similar as that in the fresh sample. Consistently, Zulueta et al. [125] did not find differences between the contents of saturated, monounsaturated, or polyunsaturated fatty acids of an untreated and a PEF-treated orange juice-milk beverage subjected to 35–40 kV/cm for 40–180 μs using 2.5-μs bipolar square-wave pulses.

12.4 Economical Considerations

Continuous operability with high flow rate capacity is recommended for an industrial implementation of PEF technology, so that the food is pumped while being exposed to the electric field at ambient or refrigerated temperatures. Therefore, the cost of the investment and the cost of running the plant are the main uncertainty when introducing a new technology in a factory. The scarce number of commercial HIPEF equipments working in an industrial scale-up level implies a high initial investment cost. At the moment, there are two industrial systems provided by DIL (Germany) and Diversified Technologies Inc. (USA), generally allowing variation of several operating parameters.

Energy and cost consumption are other of the major concerns about HIPEF technology. The equipment used, the operating regime as well as the product have a determinant influence on the final energy cost. According to Hoogland and de Haan [123], the thermal pasteurization temperature cycle can be summarized as 5-65-75-15-5°C, needing heating for 10°C and cooling for 10°C. Similarly the HIPEF temperature profile can be summarized as 5-35-50-5°C, needing heating for 30°C, HIPEF-heating 15°C and cooling for 45°C. The additional heating and cooling costs will result in higher cost for HIPEF when compared with thermal pasteurization. However, as HIPEF is nearly a non-thermal process, the cost for cleaning and down-time will be less. Faster cleaning at lower frequency than those spent by thermal process and the reduction in down-time due to less complexity and number of used parts may represent significant saving money per liter when processing a product by HIPEF (Hoogland and de Haan [123]). Therefore, in-depth studies and concept planning are important for establishing protocols for each PEF-treated product, with the ultimate goal of commercialization.

12.5 Main Research Needs, Bottlenecks, and Future Perspectives for the Industrial Implementation of PEF

Application of pulsed electric fields for food preservation offers excellent perspectives regarding the successful implementation of this environmentally friendly technology, in terms of energy consumption, at an industrial level. PEF have been shown to be an interesting technology to process acidic products such as fruit juices or even low acidic commodities, if properly combined with other processing techniques. Hence, high-quality, safe and shelf stable liquid foods can be obtained without significant depletion of their fresh bioactive potential. In-depth research is needed in order to study the factors involved in the generation/destruction of these compounds, as well as to elucidate the mechanistic insight of the changes.

However, some challenges and bottlenecks still preclude the industrial implementation of PEF. On the one hand, many groups are currently studying the effects of PEF treatments with systems that greatly differ in building-up, way of operation, and construction materials. This greatly complicates the comparison of results. On the other hand, because of the intrinsic characteristics of the technology, it is not easy to on-time monitor the process conditions that determine the boundary regions delimiting areas in which the primary treatment effects, caused by the electrical treatment, outweigh the secondary effects, caused by temperature [124]. Furthermore, industrial equipment needs to be designed considering energy recovery systems, as well as operation conditions that minimize undesirable phenomena such as electrochemical reactions leading to electrode corrosion. The development of improved PEF systems capable of adapting working conditions to different fluid characteristics and flow rates represents a significant design challenge that needs to be overcome to facilitate the commercial exploitation of this non thermal technology.

References

1. Prochownick L, Spaeth F (1890) Über die keimtötende Wirkung des galvanischen Stroms. Dtsch Med Wochenschr 26:564–565
2. Beattie JM, Lewis FC (1925) The electric current (apart from the heat generated). A bacteriological agent in the sterilization of milk and other fluids. J Hyg 24:123–137
3. Fetterman JC (1928) The electrical conductivity method of processing milk. Agr Eng 9(4):107–108

4. Moses BD (1938) Electric pasteurization of milk. Agr Eng 19(12):525–526
5. Allen M, Soike K (1966) Sterilization by electrohydraulic treatment. Science 154:155–157
6. Edebo L, Selin I (1968) The effect of the pressure shock wave and some electrical quantities in the microbicidal effect of transient electric arcs in aqueous systems. J Gen Microbiol 50:253–259
7. Gossling BS (1960) Artificial mutation of micro-organisms by electrical shock. United Kingdom, UK 845743
8. Doevenspeck H (1960) Verfahren und Vorrichtung zur Gewinnung der einzelnen Phasen aus dispersen Systemen. German Patent, DE 1,237,541
9. Doevenspeck H (1961) Influencing cells and cell walls by electrostatic impulses. Fleischwirtschaft 13(12):968–987
10. Sale AJ, Hamilton WA (1967) Effect of high electric fields on micro-organisms. I. Killing of bacteria and yeast. II. Mechanism of action of the lethal effect. BBA Gen Subj 148:781–800
11. Zimmermann U, Pilwat G, Riemann F (1974) Dielectric breakdown in cell membranes. Biophys J 14:881–899
12. Neumann E, Rosenheck K (1972) Permeability changes induced by electric impulses in vesicular membranes. J Membr Biol 10:279–290
13. Tsong TY (1991) Electroporation of cell membranes. Biophys J 60:297–306
14. Hülsheger H, Niemann EG (1980) Lethal effects of high voltage pulses on E. coli K12. Radiat Environ Biophys 18:281–288
15. Hülsheger H, Potel J, Niemann EG (1983) Electric field effects on bacteria and yeast cells. Radiat Environ Biophys 22:149–162
16. Grahl T (1994) Abtöten von Mikroorganismen mit Hilfe elektrischer Hochspannungsimpulse. TU Hamburg-Harburg, Hamburg
17. Dunn JE, Pearlman JS (1987) Methods and apparatus for extending the shelf life of fluid food products. US Patent 4,695,472, 22 Sept 1987
18. Mizuno A, Hori Y (1988) Destruction of living cells by pulsed high-voltage application. IEEE Trans Ind Appl 24(3):387–394
19. Qin BL Zhang Q Barbosa-Cánovas GV, Swanson BG (1994) Pulsed electric field pasteurization with a coaxial treatment chamber. CIFST/ICSTA 37th annual conference, P103, May 1994, Vancouver
20. Yin Y, Zhang QH, Sudhir KS (1997) High voltage pulsed electric field treatment chambers for the preservation of liquid food products. US Patent 5,690,978, 25 Nov 1997
21. Bushnell AH, Dunn IE, Clark RW, Pearlman JS (1993) High pulsed voltage systems for extending the shelf-life of pumpable food products. US Patent 5,235,905, 17 Aug 1993
22. Zhang Q, Barbosa-Cánovas GV, Swanson BG (1995) Engineering aspects of pulsed electric field pasteurization. J Food Eng 25:261–281
23. Martín O, Zhang Q, Castro AJ, Barbosa-Cdnovas GV, Swanson BC (1994) Pulse electric fields of high voltage to preserve foods. Microbiological and engineering aspects of the processing in Span. J Food Sci Technol 34:1–34
24. Palaniappan S, Richter ER, Sastry SK (1990) Effects of electricity on microorganisms: a review. J Food Process Preserv 14:393–414
25. Ho SY, Mittal GS (1996) Electroporation of cell membranes: a review. Crit Rev Biotechnol 16(4):349–362
26. Jeyamkondan S, Jayas DS, Holley RA (1999) Pulsed electric field processing of foods: a review. J Food Prot 62(9):1088–1096
27. Knorr D, Geulen M, Grahl T, Sitzmann W (1994) Food application of high electric field pulses. Trends Food Sci Technol 5:71–75
28. Mertens B, Knorr D (1992) Developments of nonthermal processes for food preservation. Food Technol 46(5):124–133
29. Barbosa-Cánovas GV, Góngora-Nieto MM, Pothakamury UR, Swanson BG (1999) Preservation of foods with pulsed electric fields. Academic, San Diego
30. Grahl T, Märkl H (1996) Killing of microorganisms by pulsed electric fields. Appl Microbiol Biotechnol 45:148–157

31. Wouters PC, Smelt JPPM (1997) Inactivation of microorganisms with pulsed electric fields: potential for food preservation. Food Biotechnol 11:193–229
32. Heinz V, Alvarez I, Angersbach A, Knorr D (2002) Preservation of liquid foods by high intensity pulsed electric fields-basic concepts for process design. Trends Food Sci Technol 12:103–111
33. Peleg M (1995) A model of microbial survival after exposure to pulsed electric fields. J Sci Food Agric 67:93–99
34. Aibara S, Esaki K (1998) Effects of high-voltage electric field treatment on bread starch. Biosci Biotechnol Biochem 62:2194–2198
35. Ho SY, Mittal GS, Cross JD (1997) Effects of high field electric pulses on the activity of selected enzymes. J Food Eng 31:69–84
36. Aguiló-Aguayo I, Soliva-Fortuny R, Martín-Belloso O (2009) Changes on viscosity and pectolytic enzymes of tomato and strawberry juices processed by high-intensity pulsed electric fields. Int J Food Sci Technol 44:2268–2277
37. Bendicho S, Barbosa-Cánovas GV, Martín O (2003) Reduction of protease activity in simulated milk ultrafiltrate by continuous flow high intensity pulsed electric field treatments. J Food Sci 68(3):952–957
38. Hodgins AM, Mittal GS, Griffiths MW (2002) Pasteurization of fresh orange juice using low-energy pulsed electrical field. J Food Sci 67(6):2294–2299
39. Odriozola-Serrano I, Bendicho-Porta S, Martín-Belloso O (2006) Comparative study on shelf life of whole milk processed by high-intensity pulsed electric field or heat treatment. J Dairy Sci 89:905–911
40. Odriozola-Serrano I, Soliva-Fortuny R, Gimeno-Añó V, Martín-Belloso O (2008) Kinetic study of anthocyanins, vitamin C, and antioxidant capacity in strawberry juices treated by high-intensity pulsed electric fields. J Agric Food Chem 56:8387–8393
41. Dörnenburg H, Knorr D (1998) Monitoring the impact of high pressure processing on the biosynthesis of plant metabolites using plant cell cultures. Trends Food Sci Technol 9:355–361
42. Toepfl S, Heinz V, Knorr D (2005) Overview of pulsed electric field processing of foods. In: Sun D-W (ed) Emerging technologies for food processing. Elsevier, Oxford, pp 67–97
43. Dunn J (2001) Pulsed electric field processing: an overview. In: Barbosa-Canovas GV, Zhang QH (eds) Pulsed electric fields in food processing: fundamental aspects and applications. Technomic Publishing Company, Lancaster, pp 1–30
44. Morris C, Brody AL, Wicker L (2007) Non-thermal food processing/preservation technologies: a review with packaging implications. Packag Technol Sci 20(4):275–286
45. IFT (2001) Effect of preservation technologies and microbiological inactivation in foods. In: IFI (ed) Evaluation and definition of potentially hazardous foods. comprehensive reviews in food science and food safety, vol 2. IFT, Chicago, pp 42–45
46. Braithwaite N, Weaver G (1990) Electronic materials. Butterworth Scientific Ltd, London.
47. Riley T, Watson A (1987) Polarography and other voltammetric methods. Wiley, NY
48. Fernandez-Díaz MD, Barsotti L, Dumacy E, Chefter JC (2000) Effects of pulsed electric fields on ovalbumin solutions and dialyzed egg white. J Agric Food Chem 48:2332–2339
49. Zimmermann U (1986) Electric breakdown, electropermeabilization and electrofusion. Rev Physiol Biochem Pharmacol 105:175–256
50. Elez-Martínez P, Escolà-Hernández J, Espachs-Barroso A, Barbosa-Cánovas GV, Martín-Belloso O (2005) Inactivation of Lactobacillus brevis in orange juice by high intensity pulsed electric fields. Food Microbiol 22(4):311–319
51. Evrendilek GA, Li S, Dantzer WR, Zhang QH (2004) Pulsed electric field processing of beer: microbial, sensory, and quality analyses. J Food Sci 69:M228–M232
52. Monfort S, Gayán E, Saldaña G, Puértolas E, Condón S, Raso J, Álvarez I (2010) Inactivation of Salmonella Typhimurium and Staphylococcus aureus by pulsed electric fields in liquid whole egg. Innovative Food Sci Emerg Technol 11:306–313

53. Vega-Mercado H, Martín-Belloso O, Qin BL, Chang FJ, Góngora-Nieto MM, Barbosa-Cánovas GV, Swanson BG (1997) Non-thermal food preservation: pulsed electric fields. Trends Food Sci Technol 8(5):151–157
54. Elez-Martínez P, Aguiló-Aguayo I, Martín-Belloso O (2006) Inactivation of orange juice peroxidase by high-intensity pulsed electric fields as influenced by processing parameters. J Sci Food Agric 86:71–81
55. Min S, Evrendilek GA, Zhang QH (2007) Pulsed electric fields: processing system, microbial and enzyme inhibition, and shelf-life extension of foods. IEEE T Plasma Sci 35:59–73
56. Alkhafaji SR, Farid M (2007) An investigation on pulsed electric fields technology using new treatment chamber design. Innovative Food Sci Emerg Technol 8:205–212
57. Min S, Jin ZT, Min SK, Yeom H, Zhang QH (2003) Commercial-scale pulsed electric field processing of orange juice. J Food Sci 68:1265–1271
58. Góngora-Nieto MM, Sepúlveda DR, Pedrow P, Barbosa-Cánovas GV, Swanson BG (2002) Food processing by pulsed electric fields: treatment delivery, inactivation level and regulatory aspects. LWT Food Sci Technol 35:375–388
59. Mosqueda-Melgar J, Elez-Martínez P, Raybaudi-Massilia RM, Martín-Belloso O (2008) Effects of pulsed electric fields on pathogenic microorganisms of major concern in fluid foods: a review. Crit Rev Food Sci Nutr 48:747–759
60. Elez-Martínez P, Soliva-Fortuny R, Martín-Belloso O (2006) Comparative study on shelf-life of orange juice processed by high intensity pulsed electric fields or heat treatment. Eur Food Res Technol 222:321–329
61. Giner J, Gimeno V, Espachs A, Elez P, Barbosa-Cánovas GV, Martín O (2000) Inhibition of tomato (Licopersicon esculentum Mill.) pectin methylesterase by pulsed electric fields. Innovative Food Sci Emerg Technol 1:57–67
62. Giner J, Gimeno V, Barbosa-Cánovas GV, Martín O (2001) Effects of pulsed electric fields processing on apple and pear polyphenoloxidases. Food Sci Technol Int 7:339–345
63. Zhong K, Wu J, Wang Z, Chen F, Liao X, Hu X et al (2007) Inactivation kinetics and secondary structural change of PEF-treated POD and PPO. Food Chem 100:115–123
64. Qin B-L, Chang F-J, Barbosa-Cánovas GV, Swanson BG (1995) Nonthermal inactivation of Saccharomyces cerevisiae in apple juice using pulsed electric fields LWT–Food Sci Technol 28:564–568
65. Elez-Martínez P, Suárez-Recio M, Martín-Belloso O (2007) Modeling the reduction of pectin methyl esterase activity in orange juice by high intensity pulsed electric fields. J Food Eng 78(1):184–193
66. Li YQ, Chen Q, Liu XH, Chen ZX (2008) Inactivation of soybean lipoxygenase in soymilk by pulsed electric fields. Food Chem 109:408–414
67. Bazhal M, Ngadi MO, Raghavan VGS (2006) Kinetics of Escherichia coli in liquid whole egg using combined PEF and thermal treatments. LWT Food Sci Technol 39:420–426
68. Elez-Martínez P, Escolà-Hernández J, Soliva-Fortuny RC, Martín-Belloso O (2004) Inactivation of Saccharomyces cerevisiae suspended in orange juice using high intensity pulsed electric fields. J Food Prot 67:2596–2602
69. El-Hag A, Otunola A, Jayaram SH, Anderson WA (2008) Reduction of microbial growth in milk by pulsed electric fields. 2008 IEEE international conference on dielectric liquids, Futuroscope-Chasseneuil, ICDL June 2008. Art. no. 4622478, pp 1–4
70. Evrendilek GA, Jin ZT, Ruhlman KT, Qiu X, Zhang QH, Ritcher ER (2000) Microbial safety and shelf-life of apple juice and cider processed by bench and pilot scale PEF systems. Innovative Food Sci Emerg Technol 1:77–86
71. Morales-de la Peña M, Salvia-Trujillo L, Rojas-Graü MA, Martín-Belloso O (2010) Impact of high intensity pulsed electric field on antioxidant properties and quality parameters of a fruit juice-soymilk beverage in chilled storage. LWT Food Sci Technol 43:872–881
72. Mosqueda-Melgar J, Raybaudi-Massilia RM, Martín-Belloso O (2007) Influence of treatment time and pulse frequency on Salmonella Enteriditis, Escherichia coli and Listeria monocytogenes populations inoculated in melon and watermelon juices treated by pulsed electric fields. Int J Food Microbiol 117:192–200

73. Mosqueda-Melgar J, Raybaudi-Massilia RM, Martín-Belloso O (2008) Non-thermal pasteurization of fruit juices by combining high-intensity pulsed electric fields with natural antimicrobials. Innovative Food Sci Emerg Technol 9:328–340
74. Mosqueda-Melgar J, Raybaudi-Massilia RM, Martín-Belloso O (2008) Inactivation of Salmonella enterica Ser. Enteritidis in tomato juice by combining of high-intensity pulsed electric fields with natural antimicrobials. J Food Sci 73:M47–M53
75. Sobrino-López A, Raybaudi-Massilia R, Martín-Belloso O (2006) High-intensity pulsed-electric field variables affecting Staphylococcus aureus in milk. J Dairy Sci 89:3739–3748
76. Yeom HW, Streaker CB, Zhang QH, Min DB (2000) Effects of pulsed electric fields on the quality of orange juice and comparison with heat pasteurization. J Agric Food Chem 48:4597–4605
77. Iu J, Mittal GS, Griffiths MW (2001) Reductions in levels of Escherichia coli O157:H7 in apple cider by pulsed electric fields. J Food Prot 64:964–969
78. Liang Z, Mittal GS, Griffiths MW (2002) Inactivation of Salmonella Typhimurium in orange juice containing antimicrobial agents by pulsed electric field. J Food Prot 65:1081–1087
79. Martín-Belloso O, Elez-Martínez P (2005) Food safety aspects of pulsed electric fields. In: Sun D-W (ed) Emerging technologies for food processing. Elsevier, Oxford, pp 183–208
80. Jayaram S, Castle GSP, Margatiris A (1992) Kinetics of sterilization of Lactobacillus brevis cells by the application of high voltage pulses. Biotechnol Bioeng 40:1412–1420
81. Sensoy I, Zhang QH, Sastry SK (1997) Inactivation kinetics of Salmonella Dublin by pulsed electric field. J Food Process Eng 20(5):367–381
82. Wouters PC, Alvarez I, Raso J (2001) Critical factors determining inactivation kinetics by pulsed electric field food processing. Trends Food Sci Technol 12:112–121
83. Aronsson K, Rönner U (2001) Influence of pH, water activity and temperature on the inactivation of Escherichia coli and Saccharomyces cerevisiae by pulsed electric fields. Innovative Food Sci Emerg Technol 2:105–112
84. Min S, Laura R, Zhang QH (2002) Water activity and the inactivation of Enterobacter cloacae inoculated in chocolate liquor and a model system by pulsed electric field treatment. J Food Process Preserv 26:323–337
85. Castro AJ, Barbosa-Cánovas GV, Swanson BG (1993) Microbial inactivation of foods by pulsed electric fields. J Food Process Preserv 17:47–73
86. Dutreux N, Notermans S, Witjez T, Góngora-Nieto MM, Barbosa-Cánovas GV, Swanson BG (2000) Pulsed electric fields inactivation of attached and free-living Escherichia coli and Listeria innocua under several conditions. Int J Food Microbiol 54:91–98
87. Mazurek B, Lubicki P, Staroniewicz Z (1995) Effect of short HV pulses on bacteria and fungi. IEEE T Dielect El In 2:418–425
88. Pothakamury UR, Vega H, Zhang Q, Barbosa-Cánovas GV, Swanson BG (1996) Effect of growth stage and processing temperature on the inactivation of E. coli by pulsed electric fields. J Food Prot 59:1167–1171
89. Qin B-L, Barbosa-Canovas GV, Swanson BG, Pedrow PD, Olsen RG (1998) Inactivating microorganisms using a pulsed electric field continuous treatment system. IEEE Transactions on Industry Applications 34:43–50
90. Barbosa-Cánovas GV, Pothakamury UR, Palou E, Swanson BG (1998) Biological effects and applications of pulsed electric fields for the preservation of foods. In: Nonthermal preservation of foods, Marcel Dekker, New York, pp 73–112
91. Marquez VO, Mittal GS, Griffiths MW (1997) Destruction and inhibition of bacterial spores by high voltage pulsed electric field. J Food Sci 62:399–401
92. Rodrigo D, Barbosa-Cánovas GV, Martínez A, Rodrigo M (2003) Pectin methyl esterase and natural microflora of fresh mixed orange juice and carrot juice treated with high intensity pulsed electric fields. J Food Prot 66:2336–2342
93. Zhang Q, Chang FJ, Barbosa-Cánovas GV, Swanson BG (1994) Inactivation of microorganisms in a semisolid model food using high voltage pulsed electric fields. LWT Food Sci Technol 27:538–543

94. McDonald CJ, Lloyd SW, Vitale MA, Perterson K, Inning F (2000) Effects of pulsed electric fields on microorganisms in orange juice using electric fields strengths of 30 and 50 kV/cm. J Food Sci 65:984–989
95. Floury J, Grosset N, Lesne E, Jeantet R (2006) Continuous processing of skim milk by a combination of pulsed electric fields and conventional heat treatments: Does a synergetic effect on microbial inactivation exist? Lait 86:203–211
96. Walkling-Ribeiro M, Noci F, Cronin DA, Lyng JG, Morgan DJ (2009) Antimicrobial effect and shelf-life extension by combined thermal and pulsed electric field treatment of milk. J Appl Microbiol 106(1):241–248
97. Sobrino-López A, Martín-Belloso O (2008) Enhancing the lethal effect of high-intensity pulsed electric field in milk by antimicrobial compounds as combined hurdles. J Dairy Sci 91:1759–1768
98. Raybaudi-Massilia RM, Mosqueda-Melgar J, Martín-Belloso O (2006) Inactivation of E. coli O157:H7, S. Enteritidis and L. monocytogenes by combining of high-intensity pulsed electric fields and malic acid in apple juice. In: Workshop on Applications of novel technologies in food and biotechnology. Cork, Ireland, 11–13 Sept 2006. Book of abstracts p 32
99. Mosqueda-Melgar J, Raybaudi-Massilia RM, Martín-Belloso O (2008) Combination of high-intensity pulsed electric fields with natural antimicrobials to inactivate pathogenic microorganisms and extend the shelf-life of melon and watermelon juices. Food Microbiol 25:479–491
100. Wouters PC, Dutreux N, Smelt JPPM, Lelieveld HLM (1999) Effects of pulsed electric fields on inactivation kinetics of Listeria innocua. Appl Environ Microbiol 65(12):5364–5371
101. Mastwijk HC Pol-Hofstad IE (2004) Pulsed electric field (PEF) processing in the fruit juice and dairy industries. Food Safety Magazine 10(3)
102. Noteborn HPJM, Lommen A, van der Jagt RC, Weseman JM (2001) Chemical fingerprinting for the evaluation of unintended secondary metabolic changes in transgenic food crops. Biotechnology 77:103–114
103. Lelieveld HLM, Wouters PC, Leon AE (2001) Pulsed electrical fields in food processing. Technomic Publishing Company, Lancaster
104. Council Directive EC (258/97), Novel food regulation, (1997)
105. Sánchez-Moreno C, Plaza L, Elez-Martínez P, De Ancos B, Martín-Belloso O, Cano MP (2005) Impact of high pressure and pulsed electric fields on bioactive compounds and antioxidant activity of orange juice in comparison with traditional thermal processing. J Agric Food Chem 53:4403–4409
106. Odriozola-Serrano I, Soliva-Fortuny R, Martín-Belloso O (2008) Changes of health-related compounds throughout cold storage of tomato juice stabilized by thermal or high intensity pulsed electric field treatments. Innovative Food Sci Emerg Technol 9:272–279
107. Odriozola-Serrano I, Soliva-Fortuny R, Hernández-Jover T, Martín-Belloso O (2009) Carotenoid and phenolic profile of tomato juices processed by high intensity pulsed electric fields compared with conventional thermal treatments. Food Chem 112:258–266
108. Odriozola-Serrano I, Soliva-Fortuny R, Martín-Belloso O (2008) Phenolic acids, flavonoids, vitamin C and antioxidant capacity of strawberry juices processed by high intensity pulsed electric fields or heat treatments. Eur Food Res Technol 228:239–248
109. Sánchez-Moreno C, Cano MP, De Ancos B, Plaza L, Olmedilla B, Granado F, Elez-Martínez P, Martín-Belloso O, Martín A (2005) Intake of Mediterranean vegetable soup treated by pulsed electric fields affects plasma vitamin C and antioxidant biomarkers in humans. Int J Food Sci Nutr 56:115–124
110. Min S, Jin ZT, Zhang QH (2003) Commercial scale pulsed electric field processing of tomato juice. J Agric Food Chem 51:3338–3344
111. Torregrosa F, Esteve MJ, Frigola A, Cortés C (2006) Ascorbic acid stability during refrigerated storage of orange-carrot juice treated by high pulsed electric field and comparison with pasteurized juice. J Food Eng 73:339–345
112. Bendicho S, Espachs A, Arántegui J, Martín O (2002) Effect of high intensity pulsed electric fields and heat treatments on vitamins of milk. J Dairy Res 69:113–123

113. Cortés C, Torregrosa F, Esteve MJ, Frígola A (2006) Carotenoid profile modification during refrigerated storage in untreated and pasteurized orange juice and orange juice treated with high-intensity pulsed electric fields. J Agric Food Chem 54:6247–6254
114. Quitão-Teixeira LJ, Odriozola-Serrano I, Soliva-Fortuny R, Mota-Ramos A, Martín-Belloso O (2009) Comparative study on antioxidant properties of carrot juice stabilised by high-intensity pulsed electric fields or heat treatments. J Sci Food Agric 89:2636–2642
115. Oms-Oliu G, Odriozola-Serano I, Soliva-Fortuny R, Martín-Belloso O (2009) Effects of high-intensity pulsed electric field processing conditions on lycopene, vitamin C and antioxidant capacity of watermelon juice. Food Chem 115:1312–1319
116. Torregrosa F, Cortés C, Esteve MJ, Frígola A (2005) Effects of high-intensity pulsed electric fields processing and conventional heat treatment on orange-carrot juice carotenoids. J Agric Food Chem 53:9519–9525
117. Zulueta A, Barba FJ, Esteve MJ, Frígola A (2010) Effects on the carotenoid pattern and vitamin A of a pulsed electric field-treated orange juice-milk beverage and behavior during storage. Eur Food Res Technol 231(4):525–534
118. Elez-Martínez P, Martín-Belloso O (2007) Effects of high intensity pulsed electric field processing conditions on vitamin C and antioxidant capacity of orange juice and gazpacho, a cold vegetable soup. Food Chem 102:201–209
119. Martín-Belloso O, Vega-Mercado H, Qin BL, Chang FJ, Barbosa-Cánovas GV (1997) Swanson, B.G. Inactivation of Escherichia coli suspended in liquid egg using pulsed electric fields. J Food Process Preserv 21:193–208
120. Li S-Q, Bomser JA, Zhang QH (2005) Effects of pulsed electric fields and heat treatment on stability and secondary structure of bovine immunoglobulin G. J Agric Food Chem 53:663–670
121. Garde-Cerdán T, Arias-Gil M, Marsellés-Fontanet R, Ancín-Azpilicueta C, Martín-Belloso O (2007) Effects of thermal and non-thermal processing treatments on fatty acids and free amino acids of grape juice. Food Control 18:473–479
122. Zulueta A, Esteve MJ, Frasquet I, Frígola A (2007) Fatty acid profile changes during orange juice-milk beverage processing by high-pulsed electric field. Eur J Lipid Sci Technol 109:25–31
123. Hoogland H, de Haan W (2007) Economic aspects of pulsed electric field treatment of food. In: Lelieveld HLM, Notermans S, de Haan SWH (eds) Food preservation by pulsed electric fields: from research to application. Woodhead publishing in Food Science, Technology and Nutrition, Cambridge, pp 257–265
124. Soliva-Fortuny R, Balasa A, Knorr D, Martín-Belloso O (2009) Effects of pulsed electric fields on bioactive compounds in foods: a review. Trends Food Sci Technol 20:544–556
125. Li Y, Chen Z (2006) Effect of high intensity pulsed electric field on the functional properties of protein isolated from soybean. Trans Chin Soc Agric Eng 22(8):194–198

Chapter 13
Emerging Risks Related to Food Technology

J. Claude Cheftel

Abstract **Global food security and safety are threatened by a number of fast-occurring changes**, even in the absence of natural disasters or terrorist attacks: overpopulation and urbanisation, environmental pollution, climate changes, intensive animal breeding, international trade and travel, emerging water- and food-borne diseases, antimicrobial-resistant bacteria, increasing food costs, complexity of food supply chains, malnutrition and risky food behaviour.

Food safety management tools, including food legislation, national and international standards, quality management systems, risk analysis, risk-based inspections and controls, monitoring and alert systems for food contaminants and food-borne diseases, quantitative microbial risk assessment, nutrition and toxicology studies, and elaborate food processing technologies have brought to consumers in developed countries a wide selection of safe foods.

Predictive and early warning and communication systems are being developed to increase the ability to "expect the unexpected" and take prevention measures before food hazards become real risks.

The production, processing, transportation, storage and/or distribution stages of modern food supply chains remain exposed to various types of biological or chemical contaminants, as evidenced by recent events or crises. The prion/BSE, dioxin, acrylamide, melamine, bisphenol A cases, and the numerous pathogen outbreaks illustrate this exposure. The melamine story and the international traffic of counterfeited foods and drinks show that profit-motivated fraud and adulteration are rising threats, opening potential paths for terrorist actions.

Recent food preservation, processing or packaging technologies and trends, in spite or because of their benefits (mild treatment, extended product shelf-life, "fresher" quality, RTE pre-cooked convenience) also bring safety risks at the consumer level: incomplete microbial inactivation, possible non respect of adequate storage conditions and expiration dates, undercooking, and generation of stress-resistant

J.C. Cheftel (✉)
Université des Sciences & Techniques, F-34000 Montpellier, France
e-mail: claude.j.cheftel@orange.fr

M. Hefnawy (ed.), *Advances in Food Protection*, NATO Science for Peace and Security Series A: Chemistry and Biology 1, DOI 10.1007/978-94-007-1100-6_13,

micro-organisms. Innovative technologies, such as the use of nanoparticules in foods or food contact materials, and the development of active, intelligent or sustainable food packaging entail uncertainties and safety concerns.

Natural disasters, droughts, floods, conflicts, and poverty often lead to emergency situations requiring large assistance operations with complex logistics and specific meals ready-to-eat or nutrient-supplemented foods. Containerised food processing units that could be deployed and quickly set to operate in production-disrupted areas are being developed by the World Food Programme. Other strategies against food insecurity include insurance policies for crop failures and renting of agricultural lands abroad.

Citizen perception of food safety risks and the EU consumers' "right to informed choice" explain why some technologies elicit rejection: ionising irradiation of foods, hormonal and antibiotic treatment of animals, the use of various "artificial" food additives, genetically modified crops and ingredients, cloned animals. Perceived benefits responding to consumers' needs (healthier, more nutritive, higher quality, more convenient, lower cost), "naturalness", respect of the environment and trusted information are the major factors influencing consumers' acceptance of innovative food technologies and products.

Novel foods and technologies are also subject to strict regulatory pre-market safety assessment and authorisation procedures. While necessary for protection against unexpected risks, some of these rules serve as barriers to innovation and trade, and fodder for strong political debates.

Keywords Food safety • Food hazard indicators • Quantitative risk assessment • Risk/benefit assessment • Rapid alert systems • RASFF • Predictive early warning systems • Globalisation • Climate changes • Intensive animal breeding • Water and soil pollution • Overpopulation • Drivers for emerging risks • Microbial pathogen outbreaks • Food supply chain • Counterfeited foods • Food bioterrorism • Fraud and adulterated foods • Food processing • Minimal processing • Under-processing • Over-processing • RTE foods • Food contaminants • Environmental chemicals • Acrylamide • Bisphenol A • Melamine • Dioxins • Mycotoxins • Hormonal disruptors • BSE • Antimicrobial-resistant micro-organisms • Stress-resistant micro-organisms • Food allergy • Obesity • Malnutrition • Radioactive contamination • Emerging technologies • Nanotechnologies • Active and intelligent packaging • Consumers' perception of risks • Novel food legislation

13.1 Introduction

Threats to food security and food safety have always existed, but modern technology and political governance have markedly improved this situation during the last decades, at least in many countries and for a large part of the world population. However, risks linked to chemical substances and biological agents often appear to be increasing, due to excessive use of industrial chemicals, pollution, overpopulation, climate changes, microbial adaptation, and resistance to antibiotics, etc. Such emerging challenges, whether natural or man-made, short- or long-term, are discussed in several recent books and reviews [1–15]. Similarly, while intentional food poisoning is not a new concept, food bioterrorism is a fairly recent one [16–26].

An emerging risk (to food safety) has been defined by EFSA, the European Food Safety Authority, as resulting from a newly identified hazard to which a significant exposure may occur, or from an unexpected increased significant exposure and/or susceptibility to a known hazard.

The present chapter first intends to review the main emerging risks to food safety (and security), their causes, their detection and alert systems, their potential prediction and prevention systems. A second part is devoted to emerging food safety risks related to food supply chains and to food processing, including novel food technologies.

13.2 Drivers for Emerging Risks in Food Safety and Security

13.2.1 Driver: Globalisation

Trade globalisation has many advantages, including year-round supply of fresh food at competitive prices, and wide commercial opportunities, with total annual exports of agricultural and food products exceeding US$ 1,000 billion worldwide. However, there are also several drawbacks:

- Growing global trade, migration and travel accelerate the spread of dangerous pathogens and contaminants in food;
- Complex international supply chains or networks with long transit distances and time increase exposure to contaminants and cross-contamination, and complicate food safety management;
- Such supply chains are vulnerable to agro/bio-terrorist attacks;
- One single contaminated food ingredient can lead to the recall of tons of food products in several countries, with high economic losses, possible import bans, damage to the tourist industry…

13.2.2 Driver: Potential for Climate Change

Although climate changes are highly complex and medium-long term phenomena with outcome also dependent on societal responses, some detrimental effects can already be identified:

- Water shortages cause quantity and quality problems with irrigation, process or ingredient water, plus possible shifts in production areas and cultured crops, and increased uses of agrochemicals;

- Some countries resort to purchases of agricultural lands abroad, while others take insurances against drought;
- Flooding may cause increased contamination of crops in the field, or increased exposure of food animals to zoonotic agents;
- Changes in temperatures and humidity are indeed expected to affect the distribution of plant and animal diseases, the production of mycotoxins, the spread of certain food pathogens, in addition to the current natural emergence of new pathogens and reservoirs of pathogens;
- Changes in the oceans and coastal environments may affect marine resources and seafood (e.g. increased algal toxins). Acidification of sea water could even hamper shell and bone formation;

13.2.3 Driver: Agriculture and Animal Breeding

While intensive agriculture and use of agro-chemicals will further impact on soil, water and energy supplies, with a high risk of contamination of water, food and feed crops by fertilisers and pesticides, the current trends in animal breeding could be even more devastating:

- Animal breeding [2] uses: ~75% of world agricultural areas (including pastures and feed crops), ~8% of the world water consumption, produces ~18% of total greenhouse gases, and has a low animal protein production yield (especially for beef);
- The huge current increase in intensive production (and consumption) of eggs, milk, and meat is threatening the livelihoods of a large number of small breeders;
- Large farms and herd size with enhanced contacts between animals, and their location near cities, increase the risk of zoonotic diseases;
- 3/4 of emerging human infectious diseases are due to animal pathogens (SARS, BSE, avian Flu, Q fever, parasitic diseases…);
- Antimicrobials given to animals as growth promoters and drugs contribute to induce antimicrobial-resistant micro-organisms.

13.2.4 Driver: Trends in Food Processing

In spite of their advantages in terms of sensorial and dietetic quality and convenience, the trends for mildly preserved foods (minimal heating, cold pasteurisation, "natural" preservatives, hurdle technology) and for lower salt or sugar contents increase the risk of surviving pathogenic micro-organisms, possibly stress-resistant, and of microbial spoilage of foods:

- Ready-to-eat products with long shelf-life enhance these risks;
- Manufacturers, including SMEs, must use appropriate product and process design, and HACCP;

- Consumers must treat such foods as perishable items requiring refrigeration and/or specific conditions of preparation;
- Refrigerated storage has created niches for psychrotrophs: *Listeria*, *Yersinia*... and noroviruses;
- Ready-to-cook products may be hazardous when undercooked (e.g. pizza in microwave ovens);
- Reduced packaging also increases microbial risks.

13.2.5 Driver: Overpopulation

Global population is predicted to rise from 6.6 billion people in 2008 to 7.4–7.8 billion in 2020, and the world food demand is being further enhanced by increasing affluent urban population in emerging countries (China, India). This should induce the following trends:

- Many countries with insufficient production capacity will become net food importers;
- Globally, due to climate changes and environmental constraints (water scarcity, risks of deforestation and loss of biodiversity…), the main sustainable solution will be to raise crop productivity on fertile land;
- Overfishing has already decreased seafood resources. Ocean drilling for oil will contribute to this decrease;
- Food prices should increase, and the global availability of certain foods decrease;
- Food choices will be affected, depending on country and income level, towards foods requiring less land and resources (e.g. less beef, more poultry and pork).

13.2.6 Driver: Demographics and Consumer Behaviour

Various current changes in consumers' demographics and concerns will impact on food safety:

- Life expectancy and the proportion of elderly people will increase, with a higher vulnerability to foodborne diseases;
- Changes in food composition motivated by consumer health (less salt, sugar or anti-microbial agents) may enhance microbial growth;
- Increased consumption of fresh, pre-packaged produce or other food eaten without heating has already caused pathogen outbreaks;
- Exotic food and ingredients, unusual animal species, small restaurants and catering, street vendors, may also enhance risks and errors;
- Organic food production and animal welfare may lead to reintroduction of pathogens with wildlife reservoirs (*Trichinella*, *Toxoplasma*) or increased prevalence (*Campylobacter*);

- Increased drug consumption due to overpopulation, urbanisation, aging, may result in drug contamination of surface and drinking waters;
- WHO estimated the number of overweight and obese persons, to increase to 2,300 and 700 million in 2015, with major health, social, and economic impacts.

13.2.7 Driver: Socio-economic Changes

Poverty, city overpopulation, economic crises, increasing food prices, compromise food security, but also food safety and quality:

- Cost is a major criterion for consumer's food selection, and manufacturers' margins of profits on food products are low, making it difficult for firms to target investment to food safety, possibly encouraging fraud;
- Food producers and processors may comply less with regulations (use sick animals, recycle foods, do not recall contaminated foods, neglect maintenance...);
- Low cost imports (e.g. from aquaculture and seafood, fresh produce) may originate from regions with poor sanitary practices or high pollution (oil spills, heavy metals, contaminated waters);
- Consumers may change consumption patterns: reduce the consumption of nutritious protein-rich animal foods, and of fresh vegetables and fruit; use foods past their expiration date;
- People with reduced immunity will be particularly vulnerable;
- Social and economic inequalities, conflicts, migrants, natural disasters, disrupt food security, requiring large emergency operations (with complex logistics) that do not solve the development problems. According to FAO, hunger affected 963 million persons in 2008, i.e. 17% of the world population (against 850 million persons in 2003–2005). In 2008, the World Food Programme has distributed 3.9 million tons of food to 102.1 million beneficiaries in 78 countries. Urgency operations concerned 25 million persons (15.7 in situations of natural disasters, 9.3 in conflict cases). Displaced persons and refugees assisted were 12.3 million. "Feeding America", USA largest domestic hunger relief organization, reports that in 2010 more than 37 million people, including 14 million children and nearly three million seniors, receive emergency food each year through the nation's network of food banks and the agencies they serve. The findings represent a 46% increase since 2006.

13.3 Search for Emerging Risks Through Recent Research Articles

A recent search for Food* and Risk* (in the titles, keywords and abstracts of articles) in the ScienceDirect Database at the beginning of 2010 revealed 534 relevant articles in 2009 and early 2010. The distribution of themes is shown below:

Main themes	Number of articles
Pathogenic micro-organisms and microbial food safety	72
Obesity and eating disorders	49
Exposure to environmental chemicals	45
Diet, lipids and cardiovascular diseases	41
Food allergy	29
Chemicals formed by food processing and storage	29
Diet and cancer	26
Food safety risk governance	20
Exposure to mycotoxins	18
Food-borne parasites and risks in animal-derived foods	17
Exposure to pesticides	15
Anti-oxidants	13
Functional foods	13
Shellfish poisoning	10
Perception of food safety risks and hygiene education	10
Under-nutrition and malnutrition	10

Other themes (with less than ten articles each) include (in decreasing order): salt, sodium and hypertension; vitamins; detrimental chemicals naturally present in foods; GMO and cloned animals; food irradiation; social determinants and impacts on food safety; viruses in foods; food environmental and sustainability challenges. Thirty-one articles concerned the effects of various diets.

13.4 Classical Food Safety Management Tools

The major food safety management tools are food legislation and standards, good agricultural practices (GAP), good hygiene practices (GHP), good manufacturing practices (GMP), HACCP [27], food quality and safety management standards (such as ISO 9001 and 22000) [28], and risk analysis [29, 30] (Fig. 13.1).

However, many newer food safety management tools have been implemented, such as the precautionary principle (in the EU legislation), independent food safety assessment agencies, third party certification, auto-controls and liability of food business operators, food traceability, national food inspections and controls (including at borders), risk-based controls, mandatory transparency and reporting of food incidents, mandatory collaboration with authorities, withdrawal and recall procedures for non-complying foods, prior notice of imported food shipments, registration of food facilities, food defence plans of individual food companies (against employees sabotage or terrorist attacks).

It is of interest to give here some details concerning recent trends in risk analysis, especially examples of **Quantitative risk assessment** (QRA).

One such example concerns the calculation of the probability of an allergic reaction (e.g. to peanut residues in a chocolate spread, when the same production line is used

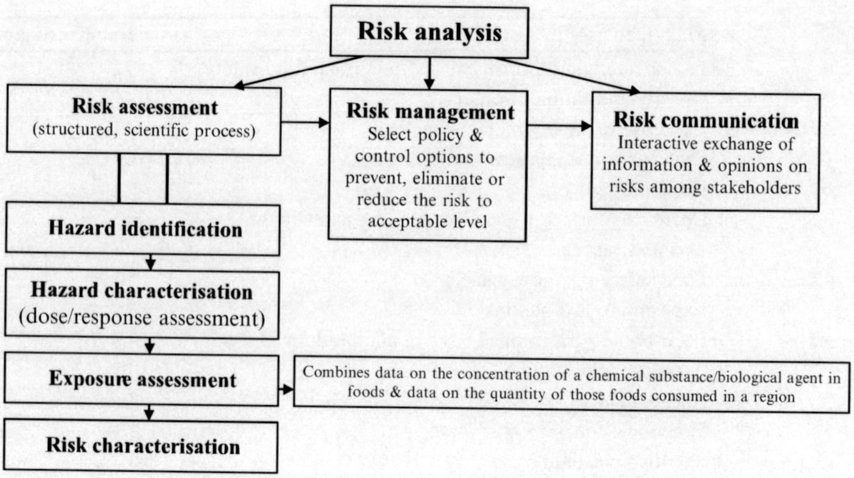

Fig. 13.1 Components of classical risk analysis. Risk is a function of probability and severity of the adverse impact of a hazard

for different products). Such a calculation would require the following estimations: (1) Proportion of the population that is allergic to the peanut allergen (perhaps 2% of adults and 4–5% of children, although there are indications that these proportions are increasing); (2) Likelihood and amount of consumption of a product (here the chocolate spread); (3) Likelihood of this food accidentally containing the allergen; (4) Concentration of allergen in this food, and minimal dose of allergen eliciting a reaction (~ 8 mg of whole peanut elicit a mild reaction in 10% of the at-risk population, as shown in a recent study in Nancy, France) [31]. A practical application would be to decide whether or not a label statement "may contain peanut" is necessary on the product (thus making it unavailable for allergic persons).

Another example deals with quantitative microbial risk assessment (QMRA): Modelling *Campylobacter* in the broiler meat production chain [32], so as (1) to assess the human illness incidence due to contaminated broiler meat; (2) to analyse the effects of control measures at different stages in the production chain.

The continuous introduction of *Campylobacter* in flocks implies that monitoring for the bacteria at the farm up to one week before slaughter may result in flocks falsely tested negative: once *Campylobacter* is established at the farm, the within-flock prevalence increases markedly within a week. Thus, at the point of slaughter, prevalence is most likely to be either very low (<5%) or very high (>95%).

All models show: (1.) a negligible effect of separate slaughter and processing of positive and negative flocks, (2.) the most effective intervention measures aim at reducing the *Campylobacter* concentration, rather than reducing the prevalence, (3.) during consumer handling, cross-contamination is more relevant than undercooking.

Another recent study from the Food Safety Authority Ireland showed that *Campylobacter* was often present on the external surface of poultry bags, and recommended the use of leak-proof packaging.

It can be noted that around 80% of chicken carcasses on the European market were found to be contaminated with *Campylobacter*, according to a region-wide baseline survey by EFSA in 2008. Prevalence of *Campylobacter* varied markedly among member states.

A third example, this time of semi-quantitative risk assessment, concerns setting priorities in monitoring antimicrobial resistance in meat and meat products [33], starting from prevalence data on antibiotic-resistant bacteria in populations of chicken, pigs, cattle, and calves.

The model estimated scores for the prevalence of antimicrobial-resistant *Campylobacter* spp., *Enterococcus* spp. and *Escherichia coli* in fresh meat, frozen meat, dried raw meat products and heat-treated meat products. The combination of the prevalence at retail, the human health impact and the amount of meat or product consumed gave the relative proportion of total risk attributed to each category of product: chicken (mostly fresh and frozen meat) contributed 6.7% of overall risk. Pork (mostly fresh meat and dried raw meat products): 4.0%. Beef and veal only 0.4% and 0.1%, respectively.

Another important and recent concept in risk analysis is that of **Risk/benefit health assessment of food.** This concept can be applied for example when the same food contains substances with adverse effects and substances with beneficial effects. The balance can shift towards benefit or risk depending on concentrations, intakes, subgroup of population. The calculation is complex [34]. As an example, one can mention the potential benefit of eating fish and seafood in adults versus the potential risk of neurotoxicity in infants and children due to contamination with methyl-mercury and/or persistent organic pollutants, from oil spills or other pollution sources (the question of sustainable fish stocks, and the possible presence of antibiotics in aquaculture products, could also be considered).

Other examples include human milk; vegetables (fibres and anti-oxidants versus anti-nutrients, pesticides, mycotoxins); nitrates-nitrites (NO homeostasis and anti-microbial effect versus nitrosamines); food fortification with vitamins and minerals (minimum recommended intakes versus risks of excess intakes).

13.5 Identification of Food Safety Risks: Rapid Alert Systems

Both HACCP and risk analysis focus on known hazards and use data related to specific hazard(s) and to given food chain(s). To make such data available to risk managers as soon as possible, public authorities have created rapid systems for identification and notification of food hazards and health risks (data should preferably be in standardised format).

The main rapid surveillance and alert systems are: (1) WHO International Food Safety Authorities Network (INFOSAN), (2) EU Rapid Alert System for Food & Feed (RASFF), (3) US Foodborne Diseases Active Surveillance Network (FoodNet) (and US PulseNet for DNA fingerprinting). These reactive systems identify issues mostly after their occurrence, and operate as communication platforms. Additional databases and networks from CDC (Center for Disease Control and Prevention, Atlanta, Georgia), ECDC (European Centre for Disease Prevention and Control, Stockholm), European Union, WHO (World Health Organization, Geneva), FAO (Food and Agriculture Organization, Rome), OIE (World Organization for Animal Health, Paris) are focused on hazards and/or on endpoints.

Concerning the World Health Organisation system (www.who.int), national contact points for emergencies both disseminate information from INFOSAN, and report to INFOSAN on hazards (contaminants in food and feed imports and exports) or endpoints (food poisonings). A Global Outbreak Alert & Response Network (GOARN) operates within INFOSAN.

The EU Rapid Alert System for Food and Feed (RASFF, http://ec.europa.eu/rasff) lists food safety hazards (presence of illegal substances or unacceptable concentrations of chemicals or pathogenic organisms) on its database, on a public weekly (and annual) basis:

https://webgate.ec.europa.eu/rasff-window/portal/index.cfm?event=notificationsList

Table 13.1 shows a list of alert notifications issued during a few days in June/July 2009.

Each Member State identifies hazards, in country or at EU borders, and must report them rapidly to RASFF, including measures taken regarding food safety (e.g. food recalls, rejection of imports not complying with food safety standards). RASSF sorts out the information and further transmits it to all Member States. Thus, RASFF acts as a notification, communication and alert platform.

In 2008, there were 1710 market notifications (alert or information), plus 1,389 border rejections (RASFF annual report 2008). An alert notification means that a decision (product withdrawal or recall…) must be taken by one or several Member States to avoid risks for consumers, while an information notification means that measures have already been taken to prevent the risks. Most notifications (in 2008) originated from official controls on the market, but some were due to a company's own checks or to consumer complaints. The distribution of alert notifications in 2008 by identified risks was as follows: (potentially) pathogenic microorganisms (24%), heavy metals (12%), mycotoxins (10%), migration (of chemicals from food contact materials) (9%), industrial contaminants (9%), allergens (6%), not determined (5%), pesticide residues (4%), residues of veterinary medicinal products (4%), foreign bodies (4%), food additives (3%), food composition (3%), GMO/novel food (2%), packaging defective/incorrect (2%), parasitic infestation (1%), microbiological contamination (1%), biotoxins (1%), biocontaminants (1%), bad or insufficient controls (1%). In 2008, there were 26 cases/notifications of food poisoning from bacteria, viruses, histamine, allergens, cocaine…, including 2 large outbreaks: (1) shellfish poisoning by precooked frozen mussels from Ireland (agent: heat-resistant azaspiracid algal toxin), (2) *Salmonella* in meat products from cooked beef steak strips from Ireland, later used by food processors (e.g. sandwich filling). Shellfish poisoning appears to be on the increase: there were outbreaks in 5 countries in March 2010, due to Norovirus in live oysters.

Kleter et al. [35] have analysed RASFF data for trends in food safety hazards during the period 2003–2007. As example of their findings, they listed the following

Table 13.1 Alert notifications

Date	Notified by	Ref.	Reason for notifying	Notification basis	Status
30/06/2009	Italy	2009.0831	Mercury (2.63 mg/kg – ppm) in smoked swordfish (Xiphias gladius) from Denmark	Official control on the market	Distribution on the market (possible)/ product (to be) seized
30/06/2009	Hungary	2009.0833	Deoxynivalenol (DON) (1464; 1613; 1398; 819; 2002; 1507; 2150; 1302 μg/kg – ppb) in durum groats from Hungary	Company's own check	Distribution on the market (possible)/ product (to be) withdrawn from the market
02/07/2009	Poland	2009.0845	Histamine (1433.5; 959.7 mg/kg – ppm) in frozen tuna steak from Spain	Official control on the market	Distribution on the market (possible)/product (to be) withdrawn from the market
02/07/2009	Germany	2009.0847	Aflatoxins (B1 = 11.3; Tot. = 11.3 μg/kg - ppb) in groundnut kernels from Sudan, via the Netherlands	Company's own check	Distribution on the market (possible)/product (to be) returned to dispatcher
02/07/2009	France	2009.0848	Diarrhoeic Shellfish Poisoning (DSP) toxins (positive) in frozen cooked mussels without shell from Germany	Official control on the market	Distribution on the market (possible)
03/07/2009	France	2009.0849	Escherichia coli O157:H7 in frozen beef from Germany, via the Netherlands	Company's own check	Distribution on the market (possible)/product (to be) returned to dispatcher
03/07/2009	Austria	2009.0850	Undeclared milk ingredient (casein: 1300 mg/kg – ppm) in dark chocolate bars from Italy	Official control on the market	Distribution on the market (possible)/product (to be) withdrawn from the market
03/07/2009	Austria	2009.0851	Undeclared milk ingredient (casein: 810 mg/kg – ppm) in soybeans, roasted in the Netherlands, coated with dark chocolate manufactured in Belgium and packaged in Austria	Official control on the market	Distribution on the market (possible)/product (to be) withdrawn from the market
03/07/2009	Finland	2009.0854	Norovirus (genogroup 2) in frozen raspberries from Poland	Food poisoning	Distribution on the market (possible)/product (to be) withdrawn from the market
03/07/2009	The Slovak Republic	2009.0858	Undeclared nuts (134.10 mg/kg – ppm) in wafers with cocoa cream from Bulgaria	Official control on the market	Distribution on the market (possible)/product (to be) relabelled
03/07/2009	The Nether lands	2009.0860	Benzo(a)pyrene (16.6 μg/kg – ppb) in food supplement from Germany	Official control on the market	Distribution on the market (possible)/product (to be) withdrawn from the market

chemical hazards (additives, contact materials, contaminants…): (1) Antibiotics: nitrofurans, chloramphenicol in seafoods, (2) Allergens: histamine, sulphites in seafoods, (3) Azo dyes: illegal Para Red and Sudan 4 in oils, condiments and spices, (4) Food contact substances: ITX ink component from packaging, in drinks, (5) Organic substances: formaldehyde, aromatic primary amines, diaminophenylmethane from Chinese plastic utensils, (6) Heavy metals: Cr, Ni, Pb, Cd from Chinese and handcraft pottery, (7) Heavy metals: Hg and Cd in seafoods, (8) Pesticides: illegal isophenphos-methyl, isocarbophos in Spanish peppers; legal pesticides (dimethoate…) above MRL in fruits and vegetables.

RASFF data for 2003–2008 were screened by Nepusz et al. [36] using algorithms to analyse levels of food alert reports against a country, and reporting countries for each alert. This monitoring tool named China, Iran, Turkey, USA, and Spain as the top five offenders in food contamination (these results are from a trading EU perspective for the period considered).

13.6 Prediction (and Prevention) of Emerging Food Safety Risks: Predictive Early Warning Systems

Previously mentioned databases and systems are essential for risk and crisis management. Their retrospective statistical analysis (frequency of occurrence of hazards and risks, thresholds, algorithms) can indicate trends. However, assessment of emerging risks requires early detection or prediction, allowing preventive measures, before they become real risks. Various attempts have been made, and systems devised, for this task of "expecting the unexpected" [37,38].

Some systems are focused on specific food hazards:

- Early warning systems for moulds and mycotoxins in maize and/or wheat: computer models based on weather variables (temperature, rainfall, relative humidity), plant development stages and agronomic criteria. Indeed climate changes affect mould growth and the level and type of mycotoxins. However these systems are not very accurate.
- Predictive microbiology: it is used to understand the growth/death/survival of micro-organisms in relation to their implicit properties and interactions, and the characteristics of the food and of the processing environment. It also predicts the shelf-life and safety of foods. Example: prediction of the risk of *Vibrio* (and phycotoxins?) in shellfish from the prediction of sea water temperatures.

Some systems are focused on food imports (automatic data screening):

- New electronic screening tool for food imports (FDA, USA): Predictive Risk-Based Evaluation for Dynamic Import Compliance Targeting (PREDICT) automatically scans many databases for information that may affect admissibility determination. For example, it will check for: A product's inherent risk rating (e.g., raw seafood); If a company or product is subject to an import alert; The compliance history of relevant companies (e.g. producer, shipper, importer) and product, including recalls, inspection results, results of field exams, sample analyses of previous entries; The admissibility history of the producer, exporter, importer,consignee; Open source intelligence about the product, producer, and foreign site from which the product originates (e.g. floods, extreme heat), entry data anomalies.

The system also looks for patterns and associations. PREDICT then issues a risk score for each entry line. Based on the score, the system will either issue a "may proceed" notice or flag the entry as potentially risky and send it to an FDA reviewer for manual processing and possible examination and sampling.

Other systems may be moderately focused using data recording plus expert or stakeholder opinion, with sharing of information and knowledge to overcome problems of misinterpretation of data and ignorance of emerging risk signals:

- ECDC DIVINE-NET project monitoring food-borne viral diseases in Europe.
- EFSA's Stakeholder Consultative Platform, with web-based posting of EFSA documents, for feedback. Activities involve stakeholders to identify or prioritise food safety issues (see www.efsa.europa.eu). EFSA, the European Food Safety Authority, in charge of risk assessment, hosts ten specialised "scientific panels" and six "cooperation and assistance units", including one on emerging risks.

Moderately focused systems assess vulnerable points in the food supply chain where the introduction of (intentional) hazards is most likely to occur:

- CARVER-shock approach (adapted and described by FDA and USDA), to shield the food supply against threats of terrorist attacks and new dangers. "Critical nodes" are identified, rather than specific hazards. Multidisciplinary experts are needed, to focus on the production chain and flow diagram for a specific sector or product.

Broadly oriented systems combine food safety with other data:

- Eurostat (EU Statistical Office) (http://epp.eurostat.ec.europa.eu/) uses its database "from farm to fork" which can show trends and developments associated with various steps in the production-consumption chain in Europe. It includes results from EU food safety monitoring programmes, numbers of food inspections carried out, trade statistics, health records.
- FAO Global Information and Early Warning System on Food & Agriculture (GIEWS) is a wide information exchange system connecting governments, NGOs, research and media institutions, in which issues related to food security (crises) are shared to help policy-makers and analysts take preventive measures. Satellite observations are used to assess agricultural land status and forecast crop harvest yields.

Broadly oriented systems (strategic foresight by experts and stakeholders) consider the position of current and desirable future scenarios, and how to reach the latter with required policy changes:

- The European Community Scientific Committee on emerging and newly identified health risks has a panel of independent experts.

- The British Department for Environment, Food and Rural Affairs (DEFRA) is involved in horizon scanning: systematic examination of potential threats, opportunities and future developments at the margins of current thinking. This may explore novel, unexpected issues and persistent problems and trends .

Other predictive systems are based on a holistic approach [12,39–41]:

- EU projects: SAFE FOODS (www.safefoods/nl/), and EMRISK

(www.efsa.europa.eu/cn/panels/emrisk.htm) which attempt to expand the scope of risk assessment, consider many sectors in and out the food production chain (science, technology, industry and innovation; international trade; nature and environment; government and policies; consumer behaviour; culture and demography; public health and welfare; agriculture; economy; information), and attempt to identify specific indicators and signals from these sectors [11].

An indicator is defined as a reliable and sensitive measurement or observation informing on the nature of the hazard and the source of risk. A signal is a change in such indicator, which may result in the development of a risk.

Additional recommendations for these holistic systems include: to increase transparency and accountability; to increase stakeholders' input (including consumers); to analyse both risks and benefits; to consider issues such as risk acceptability and mitigation.

In an article entitled "Selection of critical factors for identifying emerging food safety risks in dynamic food production chains", van Asselt et al. [42] first list the main indicators inside and outside food production chains.
Endogenous indicators:

Factors 1–6 relate to Food chain complexity: the higher possibility of errors in complex systems results in food safety risks.
Factors 7–10 are linked to Producers' attitude towards food safety.
Factors 11–13 show Producers' compliance to food safety regulations.
Factors 14–18 correspond to Technical innovations in the food chain.

Exogenous indicators

Factors 19–22 include Origin of raw materials; Legal requirements; Climate change; and Economic status.
Factors 23–29 are related to Consumers: Size of demand; Demand for convenience or quality; Consumer concern for health

Multidisciplinary experts were then asked to select critical indicators for each specific case, rank them, detect and analyse their signals. For this, group discussion was followed by individual ranking. Taking as case study a change from domestic (apple) to exotic (Indian or Australian mango) fruit chain on the Dutch market in 2000, the experts were asked to score a change in the listed indicators from –2 (much less change) to +2 (much more change). For each change in indicator, its subsequent consequences on food safety were also scored from –2 to +2 (markedly increased risk), as shown below [42].

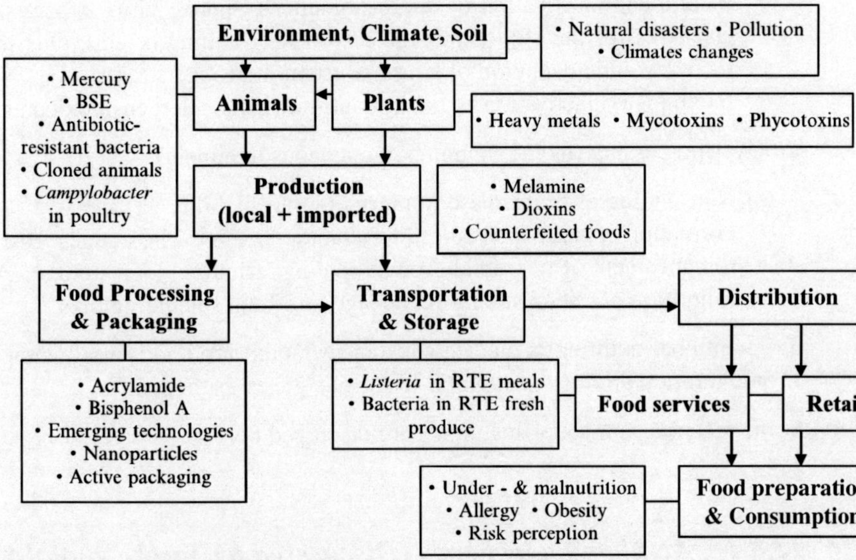

Fig. 13.2 Some safety hazards and risks in the food supply chain

Critical indicators (indicator number)	Score for change	Score for food safety risk
Logistics (5)	+2	+1
Information exchange (8)	–1	+2
Contractual agreements on quality and safety (9)	–0.5	+1
Origin of raw materials (19)	+2	+1
Legal requirements (20)	–2	+2

13.7 Safety Risks in the Food Supply Chain

Figure 13.2 summarises some of the main food safety risks in the food chain. These risks are largely due to human activities. They present a high degree of complexity and uncertainty, being related to the type of food and food attribute, type of substance, technology, region, etc.

13.8 Potential Food Safety Risks of Food Processing

They can be classified as follows:

1. Under-processing (including non homogeneity of treatment):
   - Incomplete physical removal of contaminants

- Incomplete inactivation of micro-organisms, spores, viruses, parasites (special risks for the chilled storage of foods of animal origin), plus microbial recovery and adaptation of stress-resistant microbial strains
- Incomplete inactivation of natural anti-nutrients, allergens, toxicants

2. Over-processing (including non homogeneous treatment):

- Nutrient losses or decrease in nutrient bioavailability
- Formation of toxic molecules (acrylamide, heterocyclic amines, HMF.)
- Enhancement of microbial growth (slicing, grinding, long storage…)
- Migration of chemicals from package into food (including inks…)

3. Intentional addition of unsafe substances: some food additives, nanoparticles, adulteration products, antibiotic-resistance genes, pathogens…

Some recent examples of such risks are discussed below.

13.8.1 *Acrylamide in Foods (Discovered by Tareke et al. [43])*

Acrylamide (prop-2-enamide; C_3H_5NO) may be formed in carbohydrate-rich foods: potatoes and cereal-based products (potato chips, French fries, bread, crispbreads, biscuits, breakfast cereals, coffee) during cooking processes such as frying, baking, roasting, microwaving at ≥120°C. Acrylamide levels observed were up to 5 mg/kg in potato chips and 2 mg/kg in biscuits. Acrylamide may also form in heated fat-rich dry foods such as nuts. The exact mechanisms of formation are unsure, but involve Maillard reactions between asparagine and reducing sugars/reactive carbonyls.

Acrylamide is also a highly reactive industrial chemical used for the production of plastics and materials. It may migrate into foods from some food contact materials.

Acrylamide is a known carcinogen to experimental animals, and may be a reproductive toxicant. Thus exposure from all sources, including diet, should be minimized (*Codex Alimentarius*, CAC/RCP 67-2009). Its primary metabolite, glycidamide, is also toxic.

In order to reduce acrylamide formation in foods, CIAA [44] has proposed a "Toolbox" showing how recipes and processes can be modified (use low sugar varieties and mixes, avoid overheating and brown colour, add asparaginase or amino acids, or an asparagine-degrading yeast). According to EFSA, from 2003 to 2006, the acrylamide levels of EU bread and potato chips has decreased, but not that of biscuits, breakfast cereals, French fries, or cereal-based baby foods. Further suggestions for the control of the acrylamide content of French fries are given by Sanny et al. [45].

The link between acrylamide and cancer risk in humans is uncertain. People have been exposed to acrylamide in their diet for a long time. Recent studies suggest that tolerable daily intakes of acrylamide for carcinogenic levels should be set at 2.6 μg/kg b.w., while the TDI for neurotoxicity is higher, at 40 μ/kg b.w. Recent Canadian, Swedish and US studies estimate the average exposure of adults to acrylamide in foods as 0.3, 0.5 and 0.4 μg/kg b.w. per day, respectively [46].

This example shows how even long-used processes can suddenly reveal unsuspected risks. It is also known that frying, grilling and barbecuing of meat and fish may form carcinogenic heterocyclic amines. The formation of furan or of acetaldehyde in some foods may be the next issue.

13.8.2 Hormonal Disruptors from Food Packaging and Containers

Some chemicals used as plasticizers in food packaging and/or plastic materials, and in cosmetics and adhesives, such as phtalates and bisphenol A are suspected of causing the increasing numbers of anomalies of the male sexual organs (low spermatozoid concentration, testicular cancers…) observed during the last 30 years in some industrialised European regions. Some of these chemicals, alone or in combination, exert anti-hormonal (anti-androgen or oestrogen-like) effects on animals. These chemicals could exert their adverse effects on the genital organs of male individuals, when these are exposed, orally or otherwise.

There were 46 RASFF notifications in 2008 concerning such chemicals (migration from food plastic items mainly from China). Strict rules have applied in the EU since July 2008: DEHP and DBP, suspected of being endocrine disruptors can only be used as plasticizers for articles that do not come in contact with fatty foods. Additional uses are authorised, with strict migration limits, for phtalates such as BBP, DINP and DIDP, which are not suspected of being endocrine disruptors.

Bisphenol A (BPA) (4,4′-dihydroxy-2,2-diphenylpropane) is used, usually in combination with other chemicals, to manufacture plastics and resins. BPA is thus present in polycarbonate food containers such as returnable beverage bottles, infant feeding bottles, tableware, storage containers. BPA residues are also present in epoxy resins used to make protective coatings for food and beverage cans and vats. During food storage, BPA may migrate in small amounts into food and drinks.

Concerning the safety of BPA, EFSA set a Tolerable Daily Intake of 50 μg/kg body weight in its 2007 risk assessment (TDI is the amount, on b.w. basis that can be ingested daily over a lifetime without appreciable risk). In an opinion published in 2008, EFSA noted that intakes of newborn infants, children and adults were well below the TDI. After exposure to BPA, the human body (including newborns and pregnant women) rapidly metabolises and eliminates BPA.

There are strong controversies about various possible toxic effects of BPA, and not only as hormonal disruptor (because of its binding to oestrogen receptors). BPA is permitted in food contact materials in the EU, USA, Japan, but some US states or EU countries have already banned its use for infant food containers. It is recommended not to heat polycarbonate baby bottles so as to reduce the risk of BPA migration into the food. Perhaps it could also be recommended not to leave bottles of mineral water in cars on hot days, although such containers are made of polyethylene rather than polycarbonate. Recent assessments of new studies on BPA can be found on the EFSA website.

13.8.3 Meat and Bone Flours, and the Bovine Spongiform Encephalopathy Crisis (from 1984)

Some technological changes in food or feed processing can have totally unexpected effects, that are not always completely elucidated. An example is given by the omission of technical steps (150°C fat rendering; hot solvent extraction; high temperature desolventisation) in the production process of "meat and bone flours" from ovine and bovine carcasses and offal. The simplified process probably failed to inactivate infectious prion proteins in these animal flours used as feed for cows. This may be the main cause for the spread of bovine spongiform encephalopathy (BSE) and the major resulting health, economic, consumption, and consumer trust crises.

Three other factors enhanced the BSE crisis: (1) the fact that the prion was a previously unknown infectious agent, with a "new" mode of transmission and action, (2) the recycling of infectious agents through successive animals, which appeared to increase "virulence", (3) profit-based fraud in the persistent use as feed of already banned meat and bone flours.

13.8.4 Fraud and Adulterated Foods

Profit-based criminal frauds have caused a number of food crises (through direct or indirect addition of toxic substances to foods or feeds):

1. Examples concerning **adulterated oils**:
   - Ukrainian sunflower oil containing mineral oil, exported to Europe (2008), leading to withdrawal from EU markets and temporary import ban, until strict controls were implemented (RASFF annual report 2008).
   - Oils from electric transformers fed to Belgian chicken (1999).
   - Spanish oil syndrome (1981), with 20,000 persons affected, and over 800 deaths, due to a denatured rapeseed oil (containing fatty acid esters of 3-(*N*-phenylamino)-1,2-propanediol), sold as edible olive oil.
2. Feeds and foods contaminated with **PCB and dioxins**.
3. Addition of **melamine** to gluten feed, soy meal and milk by Chinese firms or cooperative milk collection centres, to increase the nitrogen content, and mask milk solids dilution.

Dioxins

In 2008, there were 7 RASFF alert notifications for dioxins (present above legal limits) in foods, and 10 in feeds. An example is that of pork from Ireland, with a large trace and recall operation (from 54 countries) of pig meat and many processed pork products (RASFF annual report 2008). Dioxin-like PCBs and dioxins were found in pig meat at 100 fold the EU maximum limit of 1 pg/g fat. The source of this contamination/adulteration was as follows. Pigs from about 50 farms were fed with contaminated bakery wastes. These wastes consisted in bread crumbs which had been dried by direct contact with combustion gases. And the fuel burned to make the combustion gases had been contaminated by adding illegally oil from electric transformers (such oil generally contains PCB) (RASFF annual report 2008). RASFF gave detailed lists of distribution of possibly contaminated meats and meat products.

Codex alimentarius has developed a Code of practice for the prevention and reduction of dioxins and dioxin-like PCB contamination in food and feeds (CAC/RCP 62-2006).

Melamine (1,3,6-triazine-2,4,6-triamine; $C_3H_6N_6$)

Melamine is a chemical intermediate used in the manufacture of plastics and resins (and can leach into foods). According to EFSA, 2.5 mg melamine/kg food represent the critical level allowing to distinguish between unavoidable background presence and unacceptable adulteration. Intentionally contaminated Chinese foods or feeds contained up to 2600 mg/kg.

In the 2007 melamine gluten feed case, exports and incorporation into US pet foods had hurt (or killed) thousands of cats and dogs, without much impact on the Chinese food operators.

In the 2008 melamine milk case, 22 out of 79 Chinese producers of powdered infant formula were affected (and products were exported to 5 countries). The lack of transparency from Chinese authorities and the late recall caused kidney problems to over 300,000 infants and young children (over 50,000 hospitalised, at least 6 deaths). In spite of severe sanctions, melamine was still found in repackaged milk powder in Shanghai and Northern China firms in 2009–10.

Many Chinese composite products (biscuits, chocolate) containing contaminated milk powder were imported into the EU (but not milk powder as such, because importation is not permitted). Forty market notifications and 5 border rejections were listed in RASFF (RASFF annual report 2008). There were also news notifications from INFOSAN. EFSA estimated there was no health risk from these composite products, except in the case of an unlikely worst case scenario. Melamine toxicity risk data can be found in the EFSA scientific statement of 2008: the tolerable daily intake (TDI) has been reduced from 0.5 to 0.2 mg/kg b.w. Further EFSA assessments of studies on melamine exposure and toxicity can be found on the EFSA website.

13.8.5 Risks from Food Ingredients and Additives

The illegal production of foods and the use of non authorised food or feed additives are not single cases, but open and frequent practice in some countries with corruption, official silence and opaque supply and export channels.

In 2007, some 2000 Chinese producers and 1000 retailers had exported food ingredients and additives (with vitamins and amino acids) worth US$ 4 billion. Food and pharmaceutical companies worldwide used these ingredients and additives, probably without sufficient traceability or quality control. In 2010, some Chinese wheat flour was adulterated with pulverised lime added to bleaching agents (www.foodproductiondaily.com, April 9, 2010).

The Chinese Government is now reinforcing its legislation and inspection system. However, both the information and decision chains remain weak, and trained personnel is still lacking.

China is also a large market for foreign food ingredients and additives. The Food Ingredient China Exhibit (organised by an European Company) took place in Shanghai on March 23–25, 2010.

13.8.6 *Counterfeited Foods and Drinks*

The global level of fraud as counterfeiting in the food and drink industry is estimated at about $50 bn/year (www.foodproductiondaily.com, February 17 and March 24, 2009). The most frequently counterfeited foods and drinks are: fruits, conserved vegetables, baby food, milk powder, butter, instant coffee, spirits, drinks, confectionaries. Recent examples are conventionally grown vegetables sold as organic, fish sold as more premium species, canned energy drinks of unknown origin with counterfeited brand names. A picture in the French Le Monde 2 magazine (March 2010) showed the large scale destruction of counterfeited whiskies and cognacs at Zhuliao, Guangdong province, China.

Global trade and high food prices increase fraud opportunities and rewards. Counterfeiting causes high losses for food firms through damage to brands and spending on security measures. Consumers are deceived, and there is also a risk for public health.

Security measures on packaging include holograms, microdots, data codes, use-by dates, lot numbers, markers. However, counterfeiters are often able to circumvent these measures. Recent security proposals include printed pattern of fluorescent nanocrystal dots providing a fingerprint detectable under UV light, and fraud report hotlines (www.foodproductiondaily.com March 24, 2009).

13.8.7 *Likely Food Targets for Terrorism (Intentional Addition of Toxic Chemicals or Biological Agents)*

What are the most likely foods, feeds or drinks to tamper with? Those with easy access, large volumes, wide diffusion, short shelf-life, able to reach a maximum of people, or a specific group of people.

- Water: rivers, ponds, city reservoirs, tanks, sea salt ponds (although protected by water treatments, controls and dilution effect);

- Liquid foods: milk, fruit juices, beer, wine… (liquids facilitate an uniform diffusion of contaminants);
- Ingredients (possibly “foreign”) incorporated into many foods: flours, milk powder, salt, sugar, tomato paste, peanut butter, hydrolysed vegetable proteins…
- Counterfeited foods with known and trusted brands: Coca Cola, Nescafé, baby food brands;
- Food animals (and animal feeds), which can transmit zoonotic diseases to humans through consumption of animal products.

Food ingredients are special targets because they are sourced globally, at low prices, undergo less controls, are less visible and reach multiple end-products. The situation could improve if the food industry enhanced collaboration, adopting common standards, sharing supplier audits, and establishing a clearing house of suppliers. However, specification and testing usually involve compounds which are known to be a problem for a given ingredient, such as heavy metals, mycotoxins or mould, and not intentionally added toxic substances.

13.8.8 Radioactive Contamination of Foods

The main risk comes from possible accidents in nuclear power-plants, as happened in Chernobyl (Ukraine) on 26 April 1986, with considerable quantities of radionuclides released into the atmosphere, contaminating foodstuffs and feedingstuffs in several European countries to levels significant from the health point of view. Measures were adopted to ensure that certain agricultural products were only introduced into the EU according to common arrangements. A Regulation is being presently amended, laying down maximum permitted levels of radioactive contamination (in Bq/kg) of foods and feeds following a nuclear accident or any other case of radiological emergency. Such levels depend on the nature of radioelements present (Sr-90, I-131, Cs-134, Cs-137, Pu-239, Am-241…) and on the food category (infant food, dairy produce, liquid foodstuffs, other foodstuffs except “minor foodstuffs”) [47]. Because of the 30 year decay period of radioactive caesium, the EU trade in wild game, wild berries, wild mushrooms, and carnivorous lake fish of regions affected by the Chernobyl accident is still monitored and subjected to maximum permitted levels [48]. FAO, WHO and the International Atomic Energy Agency (IAEA) have also developed maximum levels of radioactivity permitted for the international trade of foods.

In a radiation emergency situation, the availability of uncontaminated food and food raw materials to consumers and to the entire production chain is a serious challenge, especially during the growing season. Hypothetical contamination scenarios and exercises are recommended for preparedness. Short-term countermeasures differ from longer-term remediation. The most frequently cited countermeasures are: food and feed restrictions, clean feeding of animals (with feed additives to bind caesium in the GI tract of ruminants), shallow ploughing of the soil, addition of ammonium ferric hexacyano-ferrate (AFCF, reduces the transfer of caesium from the soil to rye grass and clover). In all cases, measurements of radioactivity, determination of the types of radionuclides, and general information and advice to all stakeholders are required [49].

13.9 Safety of "Emerging" Food Technologies

This section starts with a brief survey of recent food technologies and their potential food safety risks. Obviously these technologies also have beneficial effects, and some of them are widely used in the food industry.

"New" physical treatments (often "minimal", or low temperature)

- High pressure pasteurisation: incomplete microbial inactivation, no spore inactivation, microbial recovery; some chemical reactions.
- Pulsed electric fields: no spore inactivation, non homogeneity and arcing; electrochemical reactions; metal transfer from electrodes.
- Cold plasma (dielectric barrier discharge at atmospheric pressure and ~30°C) (possible use for surface disinfection): free radicals, oxidations.
- Light pulses and UV: non homogeneity (surface only); photo-oxidative reactions.
- Ultrasound: low efficiency for microbial inactivation (but safe cleaning).
- Ohmic heating: under/overheating, metal transfer from electrodes.
- Microwaves: non homogeneity of heating and possible power reduction over time of domestic ovens.
- Ionising radiation: free radicals, oxidative reactions, consumer distrust (there is a pending US petition for surface irradiation of meat carcasses).

"New" chemical or combined "hurdle" treatments (often "minimal", and low temperature)

- Modified atmosphere: can delay microbial growth, but tend to over-extend shelf-life and storage time.
- Anti-microbial agents (and "hurdle" processing): incomplete microbial inactivation; microbial growth only delayed in storage; sub-lethally injured cells may give antimicrobial-resistant, acid-resistant, osmotic-resistant… or virulent strains through adaptation mechanisms; natural flora inactivation reduces competition against pathogens; potential reactivity or toxicity of some antimicrobial agents; possible use to conceal poor hygiene practices.

Examples are given by the decontamination of poultry, pork or beef carcasses by washing or spraying with solutions of: Cl_2, ClO_2, hypochlorite, acidified sodium chlorite, trisodium phosphate, organic acids (lactic, acetic), H_2O_2, O_3, electrolysed H_2O, lactoferrin… Few of these antimicrobials are presently authorised in the EU.

The efficiency and safety of some innovative anti-microbials remain to be determined:

(1) Bacteriophages against bacterial pathogens. These "bacteria-eating" viruses could be an effective way of eliminating specific food pathogens in meat and

milk products. They tend to persist longer than their hosts (replicating best on growing bacteria) and behave as inert particles in the environment. Their long term anti-bacterial activity is reduced on dry surfaces and their persistence in food varies with each bacteriophage and with the conditions of application (pH, moisture, temperature…). Refrigeration temperatures improve their persistence.

Their specificity vis à vis bacterial species, their persistence, or their ability to prevent food recontamination by pathogens is generally unknown.

In 2006, FDA accepted the GRAS status of a Listex P100 preparation of bacteriophage (obtained using *Listeria innocua*) to be used in foods to protect against *Listeria monocytogenes.*

(2) Yeasts against moulds and mycotoxins. As announced in the IFT Weekly Newsletter of Feb. 3, 2010, S. Hua from ARS-USDA found that spraying pistachio trees with the yeast *Pichia* inhibited incidence of *Aspergillus flavus* in pistachios by up to 97%, compared to unsprayed trees. The yeast can also be sprayed on the harvested or stored crop. It may also protect other crops against some microbial strains.

13.9.1 *Listeria* Risks in Ready-to-Eat Foods

The critical risk areas for this type of processed foods are:

- When these foods are targeted to vulnerable persons, such as people above 60, pregnant women and immuno-compromised persons;
- When the food composition supports the growth of *Listeria monocytogenes.* Risks are increased if pH > 4.5 and/or Aw > 0.91 (low salt content). These conditions correspond to current trends;
- Several preparation steps are also critical: the training and education of operators; cleaning operations; the slicing of meat products; packaging. The practice of HACCP is necessary;
- Current trends for minimal processing are risky for subsequent food preservation: cook-chill; moderate or cold pasteurisation, e.g. with high pressure; anti-microbial chemicals such as lactate, acetate, bacteriocins, lactoferrin, carvacrol, thymol;
- It is imperative that storage temperature be kept ≤4°C in distribution and at home. Consumers' respect of expiration date is critical. Current trends for long shelf-life increase risks.

QMRA can be used to predict efficiency of different risk mitigation options. *Codex Alimentarius* has also issued guidelines for the control of *Listeria* in RTE foods (CAC/GL 61-2007).

13.9.2 Pathogen Outbreaks in Fresh Fruit and Vegetable Produce

Fresh produce has become a frequent vehicle of foodborne illnesses (~13% of reported outbreaks with an identified food source, in the USA) [15]. This is only partly explained by the current increased consumption of fruits and vegetables, or by better outbreak detection.

Salad greens, lettuce, sprouts, and melons are leading vehicles, with norovirus, *Salmonella* and *E. coli* O157 as most frequent pathogens. The initial contamination comes through irrigation water and fertilizers. Major outbreaks are associated with fresh-cut, bagged produce (18 outbreaks in the US in 1998–2006, mostly due to leafy greens) [15]. Increased surfaces from cut, shredded, diced and/or peeled tissues release liquids and nutrients, enhance microbial attachment and growth, and interfere with disinfectant (mainly Cl_2) washes. Poorly controlled refrigerated distribution channels further increase risks.

13.9.3 Nanotechnologies

The definition of nanotechnologies is not very precise: use of substances on a very small scale (≤100 nm, obtained by assembly of molecules or downsizing). "Engineered nanomaterials" (ENM) or "nanoparticules" are already used in ingredients, additives, fertilisers, pesticides, drugs, packaging.

ENM have specific physico-chemical properties: small size increases diffusivity; high surface area enhances binding, reactivity, possibly recognition. Structure, size distribution, chemical composition, surface charge may also confer unique functionalities for food uses:

1. as food ingredients and additives: increased solubility, dispersibility, stable emulsions without emulsifier, improved texture;
2. as delivery systems for bioactive compounds: molecular traps (e.g. emulsion droplets, edible solid lipid or carbohydrate polymer particles) for protection and targeted delivery of nutrients (lycopene, phytosterols), higher availability of nutrients (Fe, Zn) for food supplements;
3. in innovative packaging: more protective or intelligent packaging: reactive nanoprobes responding to environmental changes, alerting of use-by date, temperature, pathogens, toxins; anti-bacterial coatings (e.g. silver particles) for food-contact surfaces (refrigerator…).

In plastic polymer or biopolymer films and coatings, nanoparticules may act as "fillers" to enhance barrier properties against gas migration (O_2, CO_2, H_2O vapour and/or flavours), with positive impacts on the shelf-life of fresh or processed foods. Mechanical strength and resistance to abrasion or thermal stress of films or coatings may also be improved. Nanofillers may consist of clays (montmorillonite, i.e. magnesium aluminium silicate; kaolinite), carbon-based nanotubes or graphene nanosheets.

Degradable bio-nanocomposites may also be used: starch-clay mixes, cellulose, polylactic acid, chitosan, proteins.

Current usage of nanoparticules in food/feed is increasing, but not well documented as there is no industrial inventory or public database.

There are broad uncertainties over the safe use of nanoparticules for foods and health implications of exposure [50]. Their small size increases their ability to move unexpectedly in the body. Adhesive and reactive surfaces may cause various interactions. In its Scientific Opinion of 2009, EFSA estimates that existing risk assessment methods can be applied, on a case/case basis. Data on non nano chemicals cannot be extrapolated to their nano equivalents, because formulation to the nanosize may change their properties. There are considerable limitations and uncertainties: (1) on detecting, characterising and "dosing" ENM; (2) on their absorption, distribution, metabolism, and excretion; (3) on their toxicology and environmental impact. ENM could undergo changes in the gastro-intestinal tract. Insoluble ENM may be retained and accumulate. Soluble ENM may pass through membranes, including the brain barrier, together with adsorbed substances. Little knowledge exists on their chronic exposure and carcinogenicity following oral intake. Their possible impact on the nutritional value or bioavailability of food constituents also remains to be studied. The presence of nanoparticles as contaminants in food and feed also deserves to be considered.

EFSA Scientific Opinion indicates data needed from nanotechnology applicants for risk assessment. Stakeholders have been consulted in view of future legislation. Main challenges rest with: workplace safety, distinction of natural and engineered nanoparticles, cost; food safety, uncertain legislation, including possible labelling requirements. EFSA presently (2010) prepares a guidance document on how to assess potential risks related to certain food-related uses of nanotechnology. The European Council wants "nanofoods" to be included in the future revised Novel Food Regulation.

13.9.4 Active and Intelligent Food Packaging

While "classic" packaging should be as inert as possible, these new packaging intentionally interact with the food or its environment, either ("active") to extend shelf-life with maintenance of quality, or ("intelligent") to give indication on, and monitor, the food freshness (time-temperature indicator, ripeness indicator, biosensors…). Active packaging contain deliberately incorporated components intended to release or absorb substances into or from the food or its environment (release antimicrobials, antioxidants; absorb O_2, C_2H_4…). Such packaging could be combined with tamper-proof or identity-ensuring systems. While highly promising, these new packaging are not yet widely used.

The main safety issue (as for classic packaging) is migration of chemicals and their degradation products into the food. Nano sizes could increase risks. A new Regulation (450/2009/EC) specific to active and intelligent packaging introduces safety evaluation by EFSA, with an authorisation scheme, focusing on migration data of chemicals and their toxicological properties. The efficiency of such packaging to perform the claimed function should be demonstrated in real foods. This is critical when it should prevent microbial growth, or reveal the presence of pathogenic bacteria or toxic contaminants. Their proper use should be explained to consumers by way of labelling.

The acceptance of active or intelligent packaging may be limited. Consumers may perceive systems for the extension of shelf-life as detrimental to food freshness. Time-temperature and other indicators of stressing conditions may induce consumers to select only newly displayed items. Complex packaging may also convey a negative carbon footprint image. Retailers and consumers may thus make a negative cost/benefit analysis.

Radio frequency identification (RFID) tags stuck or printed on food packages represent a type of intelligent labelling (without direct interaction with the food), replacing bar codes, that may change food sales and marketing, allowing a mobile phone to show food composition and nutritional adequacy; refrigerators to signal expiry dates and send automatic reorders; robot ovens to select cooking conditions…. RFID tags may also permit some intrusion into consumers' habits. Their reliability and resistance to tampering are not yet fully established.

13.10 Some Additional Challenges

A "Paradox of Progress" is often quoted concerning the food chain in Western Europe. In spite of renovated institutions (EFSA) and food law (Regulation EC/178/2002), stricter safety and quality standards (GMP, ISO 9001, HACCP), and intensified quality control and monitoring, the number of reported food safety incidents has increased (partly due to improved detection methods and systematic surveillance and reporting), and consumers' trust in food safety has decreased.

Several agro-food technologies tend to elicit consumer rejection (and in some cases strong political debates):

- Ionising irradiation of foods;
- Hormonal (and antibiotic) treatment of animals to hasten growth and increase meat or milk production (banned in the EU);
- Various food additives (consumers' request for "clean labels");
- Excessive use of crop fertilisers and pesticides (consumers' request for organic foods);
- Genetically modified food crops and food ingredients (in the EU);
- Genetically modified animals (including cloned animals).

There are indeed difficulties in matching the fast pace of innovation in food production and processing with risk assessment methodologies (pathogen testing, allergen testing, toxicology evaluations, environmental impact) [51].

The consumer "right to informed choice" is well established in the EU. Mandatory labelling for irradiated foods and GM foods have discouraged manufacturers and retailers to place such foods on the market, despite their potential advantages.

Consumers' perception of risks is an important challenge. Several factors influence consumer's acceptance of innovative food technologies and products [52]. Perceived benefits responding to consumers' needs (e.g. healthier, more nutritive, higher quality, lower price foods) are the major positive determinants.

Many consumers perceive new food technologies as riskier than traditional ones. "Tampering with nature" (e.g. genetic engineering) seems to be a predictor of perceived risk, while nature and naturalness are positive values. Consumers who value organic foods assess GM foods and irradiation more negatively. Chemical transformations, additives, "artificial" ingredients and gene modifications are perceived as most distant from nature, while physical transformations and genetic selection appear as less deviant.

Consumers often rely on general attitudes for judgment. Social amplification may increase the perception of risks for processes that are considered as safe by experts and policy makers. Information on risk strongly influences its perception, while familiarity with foods is more important for the perception of benefits.

Since most consumers have limited knowledge of risks or benefits of novel technologies, trust is a crucial determinant of acceptance. Consumers trust more easily operators with shared values, such as consumers' associations. Independent scientists are more trusted than national food control authorities or the food industry. Information given to consumers through labels, public debates or the media can amplify perceived risks if they do not come from a trusted source. Unintentional or deliberate misinformation often induces a fear of very low probability risks.

The technology used to create a food may dissuade consumers from buying, especially if they assume more profit to producers than to them, and if the technology deviates from naturalness. To change prior suspicious cultural and social attitudes of consumers, it is recommended to build trust, demonstrate tangible benefits, and show association with natural processes.

Legislation on novel foods and new technologies represents another issue. Depending on the category of stakeholders, it may be considered as a necessary protection against emerging food safety risks, or as a major barrier to innovation and trade.

European Regulation 97/258/EC on Novel Foods and Food Ingredients subjects each novel food to a severe pre-market safety assessment and authorisation procedure. It also specifies labelling rules to inform the consumer of any characteristics making the novel food no longer "equivalent" to an existing food, or having health or ethical implications.

According to the Regulation, there are several categories of novel foods: (1) a food not used significantly for humans in the EU before 1997; (2) a new or modified molecular structure or ingredient isolated from animals, plants, micro-organisms, fungi, or algae; (3) a food or ingredient subjected to a new process, or issued from a new production or breeding process, resulting in significant changes in composition, structure, nutritive value, metabolic effect and/or level of undesirable substances.

This classification includes some "functional foods" (which are also subject to Regulation 2006/1924/EC on Nutrition and Health Claims made on Foods). It excludes (because other rules apply) food additives, flavourings, enzymes, extraction solvents, vitamins and minerals, and (since 2003) genetically modified foods and ingredients.

Although a simplified "notification" procedure can be used for a novel food "substantially equivalent" to an existing food, Regulation 97/258/EC exerts strong constraints (costs, delays) on food business operators, detrimental to: (1) traditional foods from third countries, (2) innovative foods, (3) new technologies.

A revised European proposal (COM 2007.0872 of 14 Jan. 2008) is being discussed, maintaining high health protection, with additional objectives: (1) promote a more favourable legislative environment for innovation and competition in the food industry, (2) consider the particular needs of traditional food from third countries (avoid unjustified barriers to trade), provided these foods have a 25 years safe history of use in their country of origin. Food business operators and importers will be responsible for the safety of these foods, (3) clarify and facilitate implementation, with a single EU harmonised, centralised and shorter procedure for assessment (by EFSA) followed by authorisation (by the European Commission). Post-market monitoring by food business operators may be required, (4) give a wider choice of safe novel foods to the consumers.

13.11 Conclusions

Even in the absence of natural disasters or terrorist attacks, global food security and safety are threatened by a number of fast-occurring changes: overpopulation and urbanisation, environmental pollution, climate changes, intensive agriculture and animal breeding, international trade and travel, emerging water- and food-borne diseases, antimicrobial-resistant bacteria, increasing food costs, complexity of food supply chains, malnutrition and risky food behaviour.

Food safety management tools including food legislation, national and international standards, quality management systems, risk analysis, risk-based inspections and controls, monitoring and alert systems for food contaminants and food-borne diseases, quantitative microbial risk assessment, nutrition and toxicology studies, elaborate food processing technologies, have brought to consumers in developed countries a wide selection of safe foods.

Predictive and early warning and communication systems are being developed to increase the ability to "expect the unexpected" and take prevention measures before food hazards become real risks. Some 500 research articles listed in 2009 under food*+risk* by the ScienceDirect database also give partial indications on emerging risks.

Indeed, recent events or crises indicate that the production, processing, transportation, storage and/or distribution stages of modern food supply chains remain exposed to various types of biological or chemical contaminants. The prion/BSE, dioxin, acrylamide, melamine, bisphenol A cases, and the numerous pathogen outbreaks (*Listeria*, VTEC *E. coli*, *Campylobacter*, norovirus, parasites, shellfish toxins…) illustrate this exposure. The melamine story and the international traffic of counterfeited foods and drinks show that profit-motivated fraud and adulteration are rising threats, opening potential paths for terrorist actions.

Recent food preservation, processing or packaging technologies and trends, in spite or because of their benefits (mild treatment, extended product shelf-life, "fresher" quality, RTE pre-cooked convenience) also bring safety risks at the consumer level: incomplete microbial inactivation, possible non respect of adequate storage conditions and expiration dates, undercooking, generation of stress-resistant micro-organisms. Innovative technologies, such as the use of nanoparticules in foods or food contact materials, and the development of active, intelligent or sustainable food packaging entail uncertainties and safety concerns.

Natural disasters, conflicts, or poverty often lead to emergency situations requiring large assistance operations with complex logistics and specific meals RTE or nutrient-supplemented foods. Other strategies against food insecurity include insurance policies and renting of agricultural lands abroad.

Citizen perception of food safety risks and the EU consumer "right to informed choice" explain why some technologies elicit rejection: ionising irradiation of foods, hormonal and antibiotic treatment of animals, the use of various "artificial" food additives, genetically modified crops and ingredients. Perceived benefits responding to consumers' needs (healthier, more nutritive, higher quality, more convenient, lower cost), "naturalness", respect of the environment and trusted information are the major factors influencing consumers' acceptance of innovative food technologies and products.

Novel foods and novel technologies are also subject to strict regulatory pre-market safety assessment and authorisation procedures (European Regulations 97/258/EC on novel foods and food ingredients, and 1924/2006/EC on nutrition and health claims made on foods). While necessary for protection against unexpected risks, some of these rules are questioned in terms of barriers to innovation and trade, and objects of strong political debates.

References

1. Organisation for Economic Co-Operation and Development (2003) Emerging systemic risks in the 21st century: an agenda for action. Organisation for Economic Co-Operation and Development, Paris
2. Steinfeld H, Gerber P, Wassenaar T, Castel V, Rosales M, de Haan C (2006) Livestock's long shadow. FAO, Rome, 392 p
3. Feillet P (2007) La nourriture des Français. De la maîtrise du feu…aux années 2030. Editions Quae, Versailles
4. Bronzwaer S, Hugas M, Toldra F (eds) (2008) EFSA forum: from safe food to healthy diets. Trends Food Sci Technol 19(1): S1–S120
5. Commission of the European Communities (2008a) Future challenges for EU health and consumer policies. Directorate General for Health and Consumers, Brussels
6. Edelstein S, Gerald B, Crutchley Bushell T, Gundersen C (eds) (2008) Food and nutrition at risk in America. Food insecurity, biotechnology, food safety, and bioterrorism. Jones and Bartlett Pub, Sudbury

7. European Food Safety Authority (2008) Draft Strategic Plan 2009–2013, Parma
8. Cnossen HJ, Wassens MA, Heeres HL, Lucas Luijckx NB (eds) (2009) Vulnerabilities in the food chain. A stakeholders' guide. How to identify and address vulnerabilities. Theoretical background and perspectives. TNO Quality of Life – Sigma Chain, The Netherlands, www.sigmachain.eu/uploads/dateien/fp6-518451_stakeholders_guide_on_vulnerabilities_web.pdf
9. Dorny P, Praet N, Deckers N, Gabriel S (2009) Emerging food-borne parasites. Vet Parasitol 163:196–206
10. Institute of Food Technologists (2009) Making Decisions about the Risks of Chemicals in Foods with Limited Scientific Information. An Expert Report of IFT. Chicago
11. Kleter GA, Marvin HJP (2009) Indicators of emerging hazards and risks to food safety. Food Chem Toxicol 47:1022–1039
12. Kleter GA, Groot MJ, Poelman M, Kok EJ, Marvin HJP (2009) Timely awareness and prevention of emerging chemical and biochemical risks in foods: proposal for a strategy based on experience with recent cases. Food Chem Toxicol 47:992–1008
13. Havelaar AH, Brul S, de Jong A, de Jong R, Zwietering MH, ter Kuile BH (2010) Future challenges to microbial food safety. Intern J Food Microbiol 139(1):S79–S94
14. Quested TE, Cook PE, Gorris LGM, Cole MB (2010) Trends in technology, trade and consumption likely to impact on microbial food safety. Intern J Food Microbiol 139(1): S29–S42
15. Tauxe RV, Doyle MP, Kuchenmüller T, Schlundt J, Stein CE (2010) Evolving public health approaches to the global challenge of foodborne infections. Intern J Food Microbiol 139(1):516–528
16. WHO (2004) Public health response to biological & chemical weapons: WHO guidance. WHO, Geneva
17. WHO (2008) Terrorism threat to foods: guide for establishing or strengthening prevention and response systems. WHO, Geneva
18. Green MS, Zenilman J, Cohen D, Wiser I, Balicer RD (eds) (2005) Risk assessment and risk communication strategies in bioterrorism preparedness (NATO science for peace and security series A: chemistry & biology). Springer, New York
19. Rasco BA, Bledsoe GE (2005) Bioterrorism and food safety. CRC Press, Boca Raton
20. Pearson GS (2006) Public perception and risk communication in regard to bioterrorism against animals and plants. Rev Sci Tech Off Int Epiz 25:71–82
21. FAO (2007) FAO biosecurity toolkit. Part 1: biosecurity principles & components. Part 2: guide to assess biosecurity capacity. Part 3: an overview & framework manual for biosecurity risk analysis. Part 4: annexes. Food and Agriculture Organization, Rome
22. Commission of the European Communities (2007) Green paper on bio-preparedness. European Union, Brussels COM (2007) 399 final
23. Commission of the European Communities (2008b) Synthesis of the replies to the green paper on bio-preparedness. European Union, Brussels
24. Liu Y, Wein LM (2008) Mathematically assessing the consequences of food terrorism scenarios. J Food Sci 73:346–353
25. Maillot E (2009) Terrorisme et chaîne alimentaire. Bull Acad Vét Fr 162:27–31, http://www.academie-veterinaire-defrance.org/bulletin/pdf/2009/numero01/27.pdf
26. Turvey CG, Onyango B, Cuite C, Hallman WK (2010) Risk, fear, bird flu and terrorists: a study of risk perceptions and economics. J Socio Econ 39:1–10
27. FAO (2003) Food quality and safety systems. A training manual on food hygiene and the Hazard Analysis and Critical Control Point (HACCP) system, 3rd edn. Food and Agriculture Organization, Rome
28. Trienekens J, Zuurbier P (2008) Quality and safety standards in the food industry, developments and challenges. Int J Prod Econ 113:107–122
29. FAO (2005) Food safety risk analysis. Part I: an overview & framework manual. Part II: case studies. Food and Agriculture Organization, Rome
30. Codex Alimentarius (2007) Principles and guidelines for the conduct of microbial risk management CAC/GL63-2007

31. Taylor SL, Moneret-Vautrin DA, Crevel RWR, Sheffield D, Morisset M, Dumont P, Remington BC, Baumert JL (2010) Threshold dose for peanut: risk characterization based upon diagnostic oral challenge of a series of 286 peanut-allergic individuals. Food Chem Toxicol 48:814–819
32. Nauta M, Hill A, Rosenquist H, Brynestad S, Fetsch A, van der Logt P, Fazil A, Christensen B, Katsma E, Borck B, Havelaar A (2009) A comparison of risk assessments on *Campylobacter* in broiler meat. Int J Food Microbiol 129:107–123
33. Presi P, Stärk KDC, Stephan R, Breidenbach E, Frey J, Regula G (2009) Risk scoring for setting priorities in a monitoring of antimicrobial resistance in meat and meat products. Int J Food Microbiol 130:94–100
34. Bottex B, Dorne JLCM, Carlander D, Benford D, Przyrembel H, Heppner C, Kleiner J, Cockburn A (2008) Risk–benefit health assessment of food – food fortification and nitrate in vegetables. Trends Food Sci Technol 19(1):S113–S119
35. Kleter GA, Prandini A, Filippi L, Marvin HJP (2009) Identification of potentially emerging food safety issues by analysis of reports by the EC Rapid Alert System for Food and Feed during a 4 year period. Food Chem Toxicol 47:932–950
36. Nepusz T, Petroczi A, Naughton DP (2009) Network analytical tool for monitoring global food safety highlights China. PLoS ONE 4(8): e6680. doi:10.1371/journal.pone.0006680
37. Marvin HJP, Kleter GA, Frewer LJ, Cope S, Wentholt MTA, Rowe G (2009a) A working procedure for identifying emerging food safety issues at an early stage: implications for European and international risk management practices. Food Control 20:345–356
38. Marvin HJP, Kleter GA, Prandini A, Dekkers S, Bolton DJ (2009b) Early identification systems for emerging foodborne hazards. Food Chem Toxicol 47:915–926
39. Kuiper HA, Kleter GA (eds) (2009) Early awareness of emerging risks to food and feed safety. Food Chem Toxicol 47: 909–1040
40. Marvin HJP, Kleter GA (2009) Early awareness of emerging risks associated with food and feed production: synopsis of pertinent work carried out within the safe foods project. Food Chem Toxicol 47:911–914
41. König A, Kuiper HA, Marvin HJP, Boon PE, Busk L, Cnudde F, Cope S, Davies HV, Dreyer M, Frewer LJ, Kaiser M, Kleter GA, Knudsen I, Pascal G, Prandini A, Renn O, Smith MR, Traill BW, van der Voet H, van Trijp H et al (2010) The SAFE FOODS framework for improved risk analysis of foods. Food Control 21(12):1566–1587
42. van Asselt ED, Meuwissen MPM, van Asseldonk MAPM, Teeuw J, van der Fels-Klerx HJ (2010) Selection of critical factors for identifying emerging food safety risks in dynamic food production chains. Food Control 21:919–926
43. Tareke E, Rydberg P, Karlsson P, Eriksson S, Tornqvist M (2002) Analysis of acrylamide, a carcinogen formed in heated foodstuffs. J Agric Food Chem 50:4998–5006
44. CIAA (Confederation of the Food and Drink Industries of the EU) (2006) www.ciaa.be/asp/index.asp. The CIAA acrylamide Toolbox, Brussels
45. Sanny M, Luning PA, Marcelis WJ, Jinap S, van Boekel MAJS (2010) Impact of control behaviour on unacceptable variation in acrylamide in French fries. Trends Food Sci Technol 21:256–267
46. Tardiff RG, Gargas ML, Kirman CR, Carson ML, Sweeney LM (2010) Estimation of safe dietary intake levels of acrylamide for humans. Food Chem Toxicol 48:658–667
47. Proposal for a Council Regulation (EURATOM) laying down maximum permitted levels of radioactive contamination of foodstuffs and of feedingstuffs following a nuclear accident or any other case of radiological emergency (2010). COM (2010) 184 final. Brussels
48. Commission of the European Communities 14 April 2003 on the protection and information of the public with regard to exposure resulting from the continued radioactive caesium contamination of certain wild food products as a consequence of the accident at the Chernobyl nuclear power station (2003) Official Journal of the European Communities **L 99:** 55–56
49. Rantavaara A, Wallin H, Hasunen K, Härmälä K, Kulmala H, Latvio E, Liskola K, Mustonen I, Nieminen I, Taino R (2005) Finnish stakeholder engagement in the restoration of a radioactively contaminated food supply chain. J Environ Radioactiv 83:305–317

50. Bouwmeester H, Dekkers S, Noordam MY, Hagens WI, Bulder A, Heers C, Voorde S, Wijnhoven S, Marvin H (2009) Review of health aspects of nanotechnologies in food production. Regul Toxicol Pharmacol 53:52–62
51. Lyndhurst B (2009) An evidence review of public attitudes to emerging food technologies. Food Standards Agency, UK
52. Siegrist M (2008) Factors influencing public acceptance of innovative food technologies and products. Trends Food Sci Technol 19:603–608

Lecture Notes in Civil Engineering

Volume 154

Lecture Notes in Civil Engineering (LNCE) publishes the latest developments in Civil Engineering - quickly, informally and in top quality. Though original research reported in proceedings and post-proceedings represents the core of LNCE, edited volumes of exceptionally high quality and interest may also be considered for publication. Volumes published in LNCE embrace all aspects and subfields of, as well as new challenges in, Civil Engineering. Topics in the series include:

- Construction and Structural Mechanics
- Building Materials
- Concrete, Steel and Timber Structures
- Geotechnical Engineering
- Earthquake Engineering
- Coastal Engineering
- Ocean and Offshore Engineering; Ships and Floating Structures
- Hydraulics, Hydrology and Water Resources Engineering
- Environmental Engineering and Sustainability
- Structural Health and Monitoring
- Surveying and Geographical Information Systems
- Indoor Environments
- Transportation and Traffic
- Risk Analysis
- Safety and Security

To submit a proposal or request further information, please contact the appropriate Springer Editor:

- Pierpaolo Riva at pierpaolo.riva@springer.com (Europe and Americas);
- Swati Meherishi at swati.meherishi@springer.com (Asia - except China, and Australia, New Zealand);
- Wayne Hu at wayne.hu@springer.com (China).

All books in the series now indexed by Scopus and EI Compendex database!

More information about this series at http://www.springer.com/series/15087